T5-CVN-350

ANIMAL MODELS – DISORDERS OF
EATING BEHAVIOUR AND BODY COMPOSITION

ANIMAL MODELS – DISORDERS OF EATING BEHAVIOUR AND BODY COMPOSITION

edited by

John B. Owen
University of Wales, Bangor, UK

and

Janet L. Treasure and David A. Collier
Institute of Psychiatry, University of London, UK

KLUWER ACADEMIC PUBLISHERS
DORDRECHT / BOSTON / LONDON

A C.I.P. Catalogue record for this book is available from the Library of Congress.

ISBN 0-7923-7095-3

Published by Kluwer Academic Publishers,
P.O. Box 17, 3300 AA Dordrecht, The Netherlands.

Sold and distributed in North, Central and South America
by Kluwer Academic Publishers,
101 Philip Drive, Norwell, MA 02061, U.S.A.

In all other countries, sold and distributed
by Kluwer Academic Publishers,
P.O. Box 322, 3300 AH Dordrecht, The Netherlands.

Printed on acid-free paper

Printed in the Netherlands.

Contents

Introduction

The book aims to review knowledge on the disorders of eating behaviour and body composition in some of the non-primate higher animals and to relate these to similar conditions in humans. With advances in understanding the nature of these disorders and their biological basis, it seems timely to assess what cross-species comparisons can tell us about the general underlying factors at work. This may also help to delineate what may be a general biological basis that humans share with their higher animal comrade species and what may distinguish human from non-human, particularly in a cultural context. This could help in combating better the problems of these conditions in the animal species as well as in man and in suggesting well-based preventive measures.

As far as people are concerned the last two decades of the 20th century have shown a significant increase in obesity in the richer countries, particularly the USA (Table 1). Possibly associated with the obesity boom, there is an increasing awareness of other disorders of eating behaviour and body composition. These range from anorexia nervosa, at the other end of body composition to obesity, to others, such as bulimia, with more variable effects on body composition.

It has been apparent for some time that the recent marked change in human body condition, estimated by Body Mass Index (BMI), was not a uniform occurrence over the whole range of BMI. Although the mean value for the populations concerned was increasing there was an even more marked increase in the variance of BMI, indicating that the distribution of BMI was changing.

Flegal and Troiano (2000) have studied the distribution of BMI in the US population, recorded in the third National Health and Nutrition Examination Survey (NHANES III: 1988-94) and compared it with values from earlier comparable surveys. These were carried out for adults from 1976-1980 and for children and adolescents from 1963-1970. These data are the most comprehensive available and relate to possibly the extreme in terms of the country trends that have been examined, although other countries are fast following in the same direction. The increase in mean values of BMI between the earlier and latest surveys are shown in Table 1

These results show that for children and young people, including men up to 50 years old, the BMI values for at least the lower half of the distribution have remained remarkably constant. On the other hand the frequencies in almost all the sex/age categories have shown marked progressive increases in the upper part of the BMI distribution. Also for men in the 50+ age groups and all adult females (20+ y), except perhaps the oldest female category (70-74 y of age), there is in addition to the marked increase in frequency at the upper end of the distribution, a modest increase in BMI in the lower part of the distribution.

Table 1 Trends in mean BMI values in the USA

	Mean BMI NHANES		Prevalence of obesity ($BMI \geq 30$) NHANES	
Age-sex	II	III	II	III
Men				
20-29y	24.3	25.2	8.1	12.5
30-39y	25.6	26.5	11.9	17.2
40-49y	26.4	27.3	15.4	23.1
50-59y	26.2	27.8	14.1	28.9
60-69y	25.7	27.4	13.5	24.8
70-74y	25.4	26.8	13.4	21.1
Women				
20-29y	23.1	24.3	8.9	14.6
30-39y	24.9	26.3	16.8	25.8
40-49y	25.7	27.0	18.1	26.9
50-59y	26.5	28.4	22.4	35.6
60-69y	26.5	27.6	22.0	29.8
70-74y	26.5	27.3	18.4	25.7

There are therefore two clear components to the trend in the increased mean BMI value of the subgroups within this population. Firstly there appears to be an underlying general increase in body condition, apparent over the whole range of BMI, noticeable only in the adult women and the older male age groups. Then there is the dramatic increase in the frequencies at the upper end of the BMI distribution, noticeable in practically all age/sex groups.

This progressive increase in the population proportion at the higher BMI categories, since 1980, suggests that changes in environmental factors are involved. Of these, diet composition and level of body activity have been seen as prime causal factors. Both these factors, commonly manifest as greed and sloth, have well-known voluntary choice elements in their involvement. The suggestion of a progressive environmental causation at first sight may seem to play down the possible role of genetic factors in both eating disorders and obesity. However it is increasingly evident that there is a multifactorial causality, including the role of the environment, in uncovering thresholds of genetic susceptibility to extreme manifestations of these disorders, possibly influenced by relatively few genes. In addition there is the less pronounced progressive weight increases in some age/sex groups, over the whole range of body composition, which may indicate a more direct environmental influence.

Correspondingly trends in anorexia/weight-loss not only encompass tendencies to extremes of normal leanness but also genetic susceptibility to catastrophic breakdown in eating behaviour and resultant life-threatening emaciation. This lower end of the BMI range cannot be so clearly delineated as the upper, obese, end in these population statistics.

This is due to the low frequency of true anorexia nervosa as compared with obesity and the variable effect of the other eating disorders, on BMI.

The lack of clear understanding and delineation of the complex factors involved has undoubtedly contributed to the poor outcome of treatment and prevention measures over the whole range of these disorders, despite the enormous expenditure on them in the Western world.

The cost of these disorders in terms of human health is very high. The misery and morbidity of eating disorders are well known. Indirectly obesity can be debilitating and lead to many other serious health problems such as heart disease. Ironically our closest non-human partners, domestic pets, are also becoming prone to obesity problems.

It therefore seems important to see whether knowledge about these conditions and their causation in animal models could be a valuable asset in understanding human disorders. Observations on animals include a variety of possible relevant information, ranging from studies of diet selection in wild animals to observations and experiments on domesticated (including laboratory) animals. The use of experiments must however be delicately balanced between the need to provide positive benefits in combating difficult and sometimes intractable disorders of humans and domesticated animals and the effect on the welfare the experimental animals themselves. This is in addition to the care necessary to ensure that observations on animal models are not wrongly interpreted and applied to human therapy.

The book aims to survey some of the wealth of knowledge that is accumulating on appetite control, feeding behaviour and breakdowns in these mechanisms, resulting in a range of syndromes of obesity and emaciation, in wild and domesticated animals. These studies impinge on the general knowledge of the control mechanisms that may be common to all higher animals and are underlain by a complex genetic network. These observations can add to and enrich the store of knowledge on human subjects and provide clues for furthur research with human subjects such as that on possible underlying candidate genes.

These animal studies are even more important with the rapid disappearance of, and the paucity of archeological evidence about, the eating behaviour and body composition of anything resembling primitive hunter/gathering societies. Much of modern man's evolutionary development occurred during the operation of these primitive social structures and eating habits. Modern disorders may therefore be the consequence of recent departure, particularly in the last half-century, from more traditional lifestyles, to which Homo sapiens has not yet adapted.

Reference

Flegal KM, Troiana RP. Changes in the distribution of body mass index of adults and children in the US population. *Int J Obesity* 2000; 24: 807-18.

PART 1

Human disorders of eating behaviour and body composition

Chapter 1

OBESITY SYNDROMES

Molly S. Bray
Human Genetics Center
University of Texas Health Science Center at Houston

David B. Allison
Obesity Research Center
St. Luke's/Roosevelt Hospital &
Institute for Human Nutrition
Columbia University College of Physicians & Surgeons

ABSTRACT In this chapter we describe a variety of syndromic obesities. These are distinct obesities characterized by a specific phenotypic pattern and in many cases a single specific genetic cause. In some cases, the specific genes involved have been discovered and provide insights into both potential treatments for afflicted individuals and the mechanisms which underlie body weight regulation. The known obesity syndromes include dominant and recessive modes of inheritance, polygenic obesities and imprinted genetic transmission. Comparisons between selected animal modes of obesity and human obesity syndromes are presented. These comparisons illustrate the important cross-species effects of certain genes, point out potential candidate genes with as yet undemonstrated roles in human obesity, and highlight possible physiological pathways for further investigation in human obesity studies.

INTRODUCTION

A syndrome can be defined as a group of signs or symptoms that collectively characterize a disease or abnormal condition. To date, a number of syndromes have been identified in which obesity is a partial, if not primary, defect in the affected individual. These obesity syndromes may be the consequence of a defect in a single gene, defects in multiple genes, or may result from the complex interaction of multiple environmental, and genetic factors. Many obesity syndromes such as Prader-Willi syndrome, Bardet-

J.B. Owen et al. (eds.), Animal Models – Disorders of Eating Behaviour and Body Composition, 1–18.

Biedl syndrome, and Cohen syndrome, include mild to severe mental retardation in addition to obesity and other defects (Allison et al., 1998) suggesting that the genetic aberration producing obesity in these individuals may be of neural origin. Prevalent obesity in the population at large is associated with increased risk for insulin resistance, hypertension, type 2 diabetes mellitus, dislipidemia, and impaired glucose metabolism, and the combination of all of these chronic conditions is designated as Syndrome X, possibly the most common and complex of obesity syndromes. Because there is wide variation in the severity, prevalence, and type of inheritance pattern obesity syndromes exhibit, the underlying causes of these conditions have often been difficult to elucidate. Studies of affected individuals and their family members have provided much insight into genetic basis of the syndromes. Animal models have also served as useful tools in dissecting the functionality of the identified genetic defects but such models are often difficult to develop due to the multifactorial nature of many syndromes. Though necessarily not exhaustive, this chapter will discuss several single-gene and multigenic obesity syndromes (listed in Table 1) and the animal models that have been useful in characterizing these conditions. Further detail for selected animal models is presented in later chapters.

Table 1. Obesity Syndromes with Animal Models

Syndrome	Human Chromosome	Possible Animal Model
Simple or Single Gene Inheritance		
Leptin Deficiency	7q31	*ob* mouse
Leptin Receptor Deficiency	1p31	*db* mouse, Zucker *fatty* rat
Pro-opiomelanocortin (POMC) Deficiency	2p23	*Pomc* knockout mouse
Prohormone Convertase 1 (PC1) Deficiency	5q15-q21	*fat (Cpe)* mouse
Melanocortin Receptor 4 (MC4R) Deficiency	18q21.3	*Mc4r* knockout mouse
Single-minded 1 (SIM1) Deficiency	6q16.2	*Drosophila melanogaster* *Sim1* knockout mouse
Uncoupling Protein 3 (UCP3) Deficiency	11q13	*Ucp3* knockout mouse
Complex or Multigenic Inheritance		
Prader-Willli/Angelman Syndrome	15q11.1-q12	MatDp mouse PatDp mouse
Bardet-Biedl Syndrome/ Alstrom Syndrome	2p13-p12	*tubby* mouse
Albright Hereditary Osteodystrophy	20q13.2	*Gnas* knockout mouse

Cushing Syndrome	Polygenic	Canine, Equine, Feline, Murine
Syndrome X	Polygenic	Koletsky/SHROB rats, *fatty* rats

MECHANISMS FOR THE DEVELOPMENT OF OBESITY

Due to the complex nature of most obesity syndromes, excess body weight may result from defects in any number of pathways that regulate growth, morphological development, satiety, or energy metabolism. Much of the control of these diverse regulatory processes appears to lie within tightly localized regions of the brain, and indeed syndromic forms of obesity are often accompanied by concomitant mental retardation and/or morphological deformities. Although experimental models such as *Drosophila melanogaster* (the common fruit fly) have helped to identify genes and pathways that control structural maturation and growth, these processes are currently not well understood in humans. Nevertheless, recent research has enhanced our understanding of body weight and fat storage regulation and has supported the existence of a feedback system such as that depicted in Figure 1 (Flier, 1995) In this system, a circulating signal protein is secreted from peripheral adipose tissue and is detected by a receptor in the hypothalamus, an area of the brain that has been demonstrated to be a critical regulatory control center for feeding and metabolic rate (Kennedy, 1953; Hervey, 1959). As levels of stored body fat rise and fall, so does the concentration of the secreted, signal protein. Neurochemical and hormonal signals that stimulate or inhibit feeding and/or metabolic rate are produced from the hypothalamus in response to changes in signal protein concentrations and ultimately function to maintain an "optimal" weight for the organism (Flier, 1995; Kennedy, 1953; Harris, 1990). Thus, obesity can result from defects in any part of this feedback loop, and genes related to any of the components in this model (e.g., signal proteins, receptor proteins, neuropeptides, etc.) are candidates in which it is feasible that genetic variation could lead to disease. Genes/proteins that interact within this signaling

Figure 1. Feedback System of Body Weight Regulation

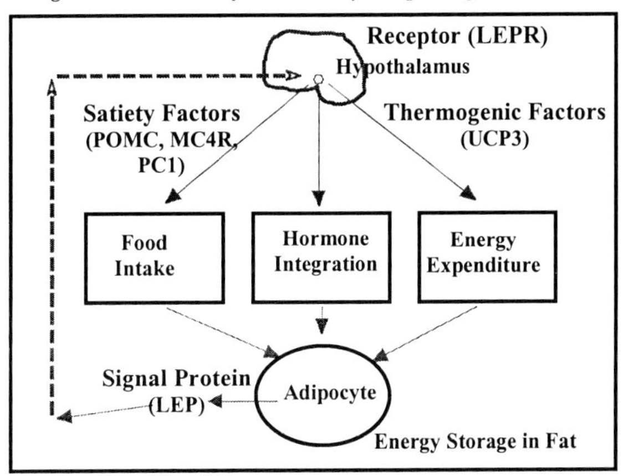

Adapted from Flier, 1995

system and produce obesity syndromes when altered in humans are noted in parentheses in Figure 1 and discussed below.

SIMPLY INHERITED OBESITY SYNDROMES

Recent years have proved to be an exciting time in the field of obesity research with the discovery of several genes responsible for obesity in a number of animal models. These animal models will be discussed in more detail in later chapters, but interestingly, human mutations in these newly discovered genes that eliminate or greatly diminish the function of their protein products also give rise to syndromic obesity phenotypes in humans very similar to that of the animal models. Nearly all of the genes identified in these animal and human models of obesity appear to be major components of the signaling system described above and produce obesity both through unregulated feeding as well as impaired energy metabolism.

Leptin Deficiency

In 1950, the *obese (ob)* mutation occurred spontaneously in a mouse colony at the Jackson Laboratory and produced a phenotype that included gross obesity coupled with features of type 2 diabetes (Ingalls et al.,1950). After many years of characterizing and analyzing the genetics of the *ob* mutant mice, Friedman and colleagues reported the discovery of the leptin (LEP) gene, heralded as a breakthrough in obesity research, in late 1994 (Zhang et al., 1994). Many reports have provided support for the role of leptin as the long hypothesized signal of stored adipose to the brain. Numerous studies have shown that leptin levels correlate to the amount of stored body fat in both humans and animals, that infusion of leptin decreases feeding and increases metabolic rate in rodents, and that leptin levels rise and falls concurrently with changes in body fat stores. (Considine et al., 1996; Frederich et al., 1995; Halaas et al., 1995; Maffei et al., 1995; Pellymounter et al., 1995; Lonnqvist et al., 1995; Hamilton et al., 1995). Nevertheless, leptin does not appear to serve simply as an "energy sensor" but has also been shown to influence both lipid and glucose metabolism, and to interact with insulin signaling pathways as well (Cohen et al., 1996; Mueller et al., 1997; Muoio et al., 1997; Chen et al., 1996; Collins et al., 1996).

Despite the physiological evidence that leptin appears to be a central factor in the regulation of body fat stores, no common variation in the LEP gene has been shown to be associated with obesity or body size measures in the general population. Nevertheless, complete absence of the leptin protein produces obesity syndromes in humans very similar to the rodent models with similar defects. Montague et al. reported the first identified case of leptin deficiency due to the presence of a rare recessive LEP mutation in two morbidly obese children from a large inbred Pakistani kindred. The mutation results in deficient production of leptin protein, pronounced hyperphagia (overeating), and elevated insulin levels supporting the contention that leptin is a key regulator of energy balance in humans (Montague et al., 1997). Subsequently, a separate mutation that effectively eliminates all functional leptin protein was observed in three individuals in a consanguineous Turkish pedigree (Strobel et al., 1998). This mutation produces morbid obesity, hypoleptinemia, hyperphagia, hyperinsulinemia, and hyperglycemia. All other family members who were either heterozygous or homozygous for the normal allele had normal weight and normal leptin, insulin, and fasting blood glucose levels. These human mutations mimic that of *ob* mice, in that inheritance is recessive (i.e., two mutated copies of the gene are necessary to produce disease), and production of functional leptin protein is completely abolished resulting in syndromic obesity in humans.

Leptin Receptor Deficiency

Shortly after the discovery of leptin, the leptin receptor (LEPR) gene was identified by several groups, and shown to be expressed in the hypothalamus and areas of the brain proximal to the hypothalamus (Lee G-H et al., 1996; Tartaglia et al., 1995). Several rodent models of obesity, including C57BL6 *db/db* mice, Zucker and Wistar *fa/fa* rats and Koletsky spontaneously hypertensive *f/f* rats have since been demonstrated to have mutations within the LEPR gene (Lee G-H et al., 1996; Takaya et al., 1996; Chua et al., 1996; Iida et al., 1996; Phillips et al., 1996; Wu-Peng et al., 1997). Each strain possesses a different mutation within LEPR that essentially inactivates the receptor protein and produces a phenotype that includes gross obesity, hyperphagia, hyperinsulinemia, hyperlipidemia, insulin resistance, severe diabetes (*db/db*), and hypertension (*fa/fa* and *f/f*) (Coleman, 1978; Kurtz et al., 1989). The discovery that all three rodent models have mutations within the same gene suggests that leptin-receptor binding must induce a multiplicity of genetic effects and that disruption of LEPR results in general syndromic perturbation of several interrelated metabolic processes.

As might be predicted, leptin receptor deficiency produces a very similar phenotype as that of leptin deficiency. Though a great deal of DNA sequence variation has been observed in the LEPR gene, no single common variant has been consistently associated with obesity (Heo et al., 2001) with the exception of a rare recessive mutation in LEPR reported by Clement and colleagues in a consanguineous family that produces a truncated receptor protein and results in uncontrolled feeding and morbid obesity within the first months of life (Clement et al., 1998). This defect also produces social impairment and emotional lability but no overt mental retardation in the affected individuals. Very recently, an obese three-year-old child weighing 59.5 kg was identified as having a defect in the leptin receptor, resulting in voracious overeating and uncontrolled weight gain (Roybal, 2000). In syndromes resulting from defects in the leptin and leptin receptor, obesity is primarily the result of overeating, as opposed to depressed metabolic rate, and all adults demonstrate impaired reproductive function and hypogonadism, suggesting that leptin also plays a role in pubertal development.

Melanocortin Pathway Defects

Pro-opiomelanocortin (POMC) is a precursor protein produced in the brain and other areas of the body whose cleavage products are critical in the regulation of pigmentation, adrenal function, pain perception, analgesia, exocrine gland function, and body weight (Krude and Gruters, 2000). Processing of POMC is accomplished by the enzymatic action of prohormone convertase 1 (PC1), which cleaves the precursor protein into adrenocorticotropin (ACTH), melanocortins (including α-melanocyte-stimulating hormone, α-MSH), and β-endorphin. One of the oldest animal models of obesity is the *agouti* mouse, which has distinctive yellow coat hair and is markedly obese. The *agouti* protein is normally produced in neonatal skin cells where it inhibits binding of α-MSH to melanocortin-1 receptors and switches the pigmentation pathway of skin from eumelanin (black) to pheomelanin (yellow). A chromosomal rearrangement in the *agouti* mutant mouse causes the agouti protein to be produced ubiquitously throughout the body (Miller et al., 1993). Within the brain, α-MSH normally acts to control body weight by binding to the melanocortin-4 receptor (MC4R) and inhibiting food intake. When agouti protein is produced in the brain, it prevents the binding of α-MSH to MC4R and induces uncontrolled overeating and adult-onset obesity in *agouti* mice. A naturally occurring neural homologue to *agouti*, agouti related protein (*AgRP*), has been shown to be elevated in *obese* mice and to produce obesity in transgenic mice that overexpress the protein (Shutter et al., 1997; Graham et al., 1997). The discovery of AGRP and the observation that *POMC* is expressed in regions of the brain that also express the leptin receptor have identified the melanocortin system as one pathway with which the leptin signaling system may interact to regulate fat storage and utilization. A

Figure 2. Melanocortin-Leptin Interaction

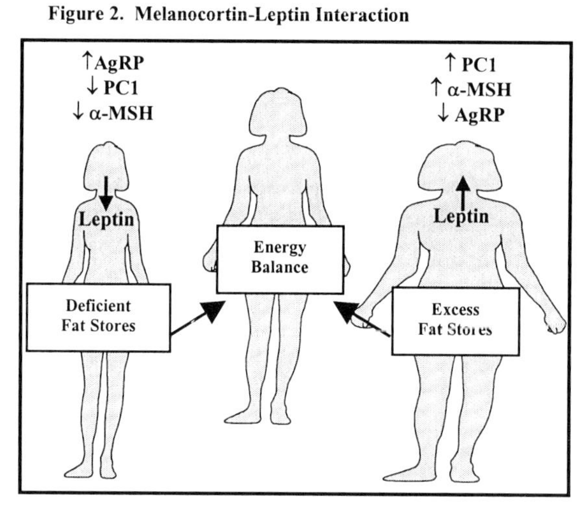

schematic of leptin-melanocortin interaction in depicted in Figure 2. In this model, depletion of fat stores (figure on left) accompanied by a concomitant decrease in leptin levels stimulates production of AgRP and inhibits the action of PC1 and α-MSH in the brain in order to stimulate feeding and restore energy balance. Conversely, in the case of excess fat stores (figure on right), increasing leptin levels activate PC1 cleavage of POMC to form α-MSH and inhibit production of AgRP resulting in decreased feeding and again a return to energy balance. Defects in *POMC*, *PC1*, and *MC4R* have each been shown to produce obesity in humans.

POMC Deficiency Syndrome

Krude et al. reported the first human mutations within the *POMC* gene. These mutations, which essentially eliminate all functional POMC protein in the affected individuals, produce both red hair pigmentation and extreme obesity (Clement et al., 1998). ACTH is completely absent in all affected patients, giving rise to hypoglycemia and disorders of bile processing. A first-born boy, who died at the age of 7 months due to hepatic failure resulting from undiagnosed adrenal insufficiency, was later found to have compound heterozygosity of two mutations in *POMC* (i.e., each of the two copies of his gene contained a different mutation, one from the father and one from the mother), both of which produced a truncated, nonfunctional protein. His sister had the same two *POMC* mutations and was treated with hydrocortisone substitution shortly after birth, preventing the deleterious effects of cortisol deficiency seen in her brother. Nevertheless, while mental development was normal, she had an increased appetite beginning at 4 months of age giving rise to severe obesity that interfered with her ability to walk until she was 2 years old (Krude et al., 1998). Three unrelated pediatric patients from different populations were found to have mutations in the non-coding region of exon 2 that eliminated translation of the POMC protein and also produced early-onset obesity resulting from uncontrolled eating (Krude and Gruters, 2000). All parents with a single *POMC* mutation are unaffected by either obesity or cortisol deficiency and none have red hair, suggesting that inheritance of this syndrome is recessive. *Pomc* knockout mice also develop hyperphagia, severe obesity, and altered pigmentation similar to humans with defects in *POMC* (Yaswen et al., 1999).

Prohormone Convertase 1 Deficiency

Activation of the POMC precursor protein requires enzymatic cleavage by PC1, and deficiency in PC1 has been shown to produce syndromic obesity very similar to that of POMC deficiency syndrome. O'Rahilly et al. (1995) first described an adult patient with postprandial hypoglycemia, hypogonadism, cortisol deficiency, elevated plasma proinsulin and POMC, and severe early-onset obesity. The patient weighed 36 kg at the age of 3 years but was later treated effectively with dietary intervention (O'Rahilly et al., 1995). The increased levels of the precursor hormones proinsulin and POMC were suggestive of a defect in prohormone processing, and this patient was later found to have compound heterozygosity for two mutations in *PC1* that eliminated production of all functional PC1 enzyme. A mouse model of obesity, the *fat* mouse, has a defect in the gene encoding carboxypeptidase E (*Cpe*), an enzyme with very similar actions to those of PC1. The *fat* mouse presents a phenotype of obesity, diabetes, infertility, and defective processing of both proinsulin and POMC (Naggert et al., 1995). Thus, the *fat* mutation was the first example of an obesity-diabetes syndrome elicited by a genetic defect in a prohormone processing pathway.

Melanocortin Obesity Syndrome

Animal models in which the *Mc4r* gene has been completely inactivated are hyperphagic, hyperinsulinemic, have increased linear growth, and develop maturity-onset obesity very similar to that of *agouti* mutant mice (Huszar et al., 1997). Interestingly, mice with only one inactivated copy of the *Mc4r* gene have adipose stores intermediate between normal and homozygous knockout mice, suggesting that haploinsufficiency of the receptor protein is sufficient to cause perturbations in energy balance. Recent studies of the *MC4R* gene in humans have shown it to be highly variable with almost thirty different variants reported to date. Like *Mc4r* heterozygous knockout mice, a single deleterious copy of the *MC4R* gene can produce obesity in humans. Vaisse et al. (1998) and Yeo et al. (1998) identified the first two families with mutations in the *MC4R* gene by screening two different samples of morbidly obese individuals. A single obese individual in each study was found to have a mutation in the *MC4R* gene that produced a truncated receptor protein lacking domains critical for receptor binding and signaling. Affected individuals had normal adrenal function and sexual development and greater than average height, consistent with *Mc4r* knockout mice. Further investigation demonstrated that severe obesity was transmitted through each generation of the subjects' respective families concurrently with the *MC4R* mutations, and the mutations were not found in any normal-weight individuals (Vaisse et al., 1998; Yeo et al., 1998). Subsequent to these early reports, several studies have since reported a number of mutations in the *MC4R* gene leading to obesity (Hinney et al., 1999). Of these mutations, many have been shown to produce alterations in binding affinity or receptor activation and most are associated with extreme obesity and overeating resulting from haploinsufficiency of the receptor protein (Vaisse et al., 2000). Though almost all carriers of *MC4R* mutations are obese, age of onset and severity of obesity is variable and there are isolated cases of carriers who are not obese (Gu et al., 1999), suggesting that other factors may mediate the effects of these mutations.

Other Single Gene Obesity Syndromes

In addition to the syndromes listed above, two additional single-gene variants have been reported to produce obesity in humans. Recently, a five-year-old girl with profound obesity who weighed 47.5 kg and demonstrated an aggressive, voracious appetite was identified as having a chromosomal rearrangement resulting in fusion of the short arm of chromosome 1 to the long arm of chromosome 6 (Holder Jr et al., 2000). Molecular studies indicated that the chromosomal disruption lay within the *SIM1* gene. The *SIM1* gene encodes a human homologue of the *Drosophila* transcription factor called *Sim* (single-minded), which has been demonstrated to be critical in neural development. Though not implicated in regulatory pathways of body weight prior to this point, *SIM1* may play a role in transcription of signaling molecules important for neural regulation of body fat stores. In addition to *SIM1*, a mutation in the uncoupling protein 3 gene (*UCP3*) that replaces a tryptophan with an arginine at codon 70 has been reported in a 15-year-old Chinese male with severe obesity and type 2 diabetes (Brown et al., 1999). *UCP3* is an important regulator of energy metabolism and body temperature. Stimulation of this protein increases utilization of stored energy, and defects in this protein result in decreases in both metabolic rate and lipolysis. The mutated protein showed completed loss of activity when expressed in yeast cells.

COMPLEX OR MULTIGENIC OBESITY SYNDROMES

Currently, there are over 40 different complex syndromes listed in the Online Mendelian Inheritance in Man, an electronic compilation of genetic disorders, which include obesity as one component of the clinical diagnosis (OMIM (TM), 2000). Multigenic obesity syndromes in which the genetic abnormalities have been identified or localized to specific chromosomal regions include Prader-Willi Syndrome, Angelman Syndrome, Bardet-Biedl Syndrome, Albright Hereditary Osteodystrophy, Ulnar-Mammary Syndrome, Borjeson-Frossman-Lehmann syndrome, Cohen Syndrome and Alstrom syndrome among others (Gunay-Aygun et al., 1997). Family studies have been useful in localizing the genetic defects for these inherited syndromes, and animal models have helped in determining how these genetic defects functionally alter the development and physiology of the affected individuals. Representative examples of multigenic obesity syndromes are presented below.

Prader-Willi Syndrome

Prader-Willi syndrome (PWS) is one of the most common obesity syndromes, with an estimated prevalence of 1/10,000 to 1/15,000 (Gunay-Aygun et al., 1997). PWS is characterized by fetal and infantile inactivity, failure to thrive in infancy, deficient sexual development, infertility, early-onset hyperphagia producing severe childhood obesity, short stature, small hands and feet, mild mental retardation, and severe learning disabilities (Cassidy,1984). PWS and another syndrome, Angelman syndrome (AS) result from disorders of imprinted genes in chromosomal region 15q11-q13. Though AS does not include obesity as part of its primary clinical diagnosis and has distinct features from those of PWS (severe mental retardation, tremulousness, uncoordinated gait, seizures, hyperactivity, cheerful demeanor, and inappropriate laughter) (Williams et al., 1995), the mechanism of inheritance is similar and will provide a good example of the consequences of loss of function of imprinted genes. Imprinting refers to the selective silencing of one copy of a gene or genes based on its parent of origin (Barlow, 1995). Certain genes are "tagged," possibly via DNA methylation, to denote from which parent they were inherited, and imprinting allows for certain traits to be expressed selectively from the maternal or paternal genes. Nevertheless, the loss of imprinted genes can produce differential and deleterious effects as seen in PWS and AS. Most commonly, PWS and AS result from chromosomal deletions of 15q11-q13, but all PWS deletions are from the paternal chromosome and all AS deletions come from the maternal chromosome, illustrating that manifestation of disease is dependent on the origin of the defect and that both maternal and paternal copies are necessary for normal development (Gunay-Aygun et al., 1997). Rather than a paternal deletion of 15q11-q13, about 25% of PWS patients have two copies of the maternal chromosome, rather than one from each parent, known as uniparental disomy. Molecular studies in mouse models of PWS have indicated that the critical region on chromosome 15 is likely to be contain several genes that influence PWS. Cattanach et al. (1997) have developed two mice models with either maternal (MatDp) or paternal (PatDp) duplication of mouse chromosome 7, which has homology to human chromosome 15. MatDp mice die shortly after birth due to an inability to feed,

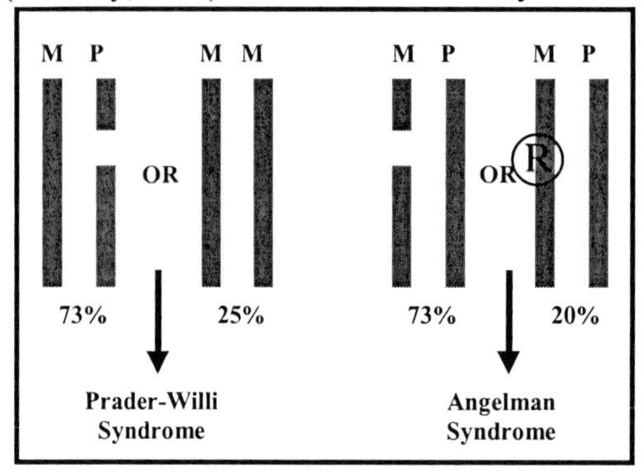

Figure 3. Maternal and Paternal Inheritance Patterns of Prader-Willi and Angelman Syndromes. The majority of Prader-Willi and Angelman patients (73%) have deletions in either the paternal (PWS) or maternal (AS) chromosome 15. The next most common class of patients have two copies of the maternal chromosome and no paternal chromosome (PWS) or a single defect in the maternal chromosome (AS).

similar to individuals with PWS. A transgenic mouse model of PWS and AS (TgPWS/AS(del)) has the homologous region to the chromosome 15q11-q13 completely eliminated. Paternal transmission of the deletion produces PWS, while maternal transmission produces AS in these mouse models. As no patient with PWS has been shown to have a mutation that affects expression of only one gene, PWS is hypothesized to result from the loss of expression of several genes within the deletion sequence. Studies of mouse models and humans have implicated several genes in the etiology of PWS. The SNRPN gene, thought to influence alternative splicing of transcripts in the brain and the development of nervous system function, was the first gene to be identified in the PWS deletion region (Kuslich et al., 1999). Dittrich et al. (1996) demonstrated that an imprinting center lay within the PWS deletion region on chromosome 15 and that this imprinting center encodes alternative transcripts of the SNRPN gene. Exons lacking protein coding potential are expressed from the paternal chromosome only. More recently, the necdin (*NDN*) gene has been mapped to the PWS deletion region and been shown to be expressed exclusively from the paternal allele in mice and in human fibroblasts. *NDN* is thought to regulate neuronal growth and nervous system development and loss of expression may be related to the mental defects and obesity observed in PWS patients (MacDonald and Wevrick, 1997).

Bardet-Biedl Syndrome/Alstrom Syndrome

Bardet-Biedl syndrome (BBS) is characterized by retinitis pigmentosa, obesity, mental retardation, dysmorphic hands and feet, hypogenitalism, and renal dysfunction. Despite the fact that the same symptoms present across all Bardet-Biedl patients, genetic studies of large pedigrees with multiple affected individuals have identified at least five different genomic regions that appear to be involved in BBS. Among several different populations, four potential BBS genes have been localized to specific chromosomal regions, including 11q13 (BBS1) (Leppert et al., 1994), 16q21 (BBS2) (Kwitek-Black et al., 1993), 3p13-p12 (BBS3) (Sheffield et al., 1994), and 15q22.3-q23 Carmi et al., 1995). A fifth BBS gene (BBS5) has recently been implicated in a study of an isolated population from Newfoundland, in which no previously identified BBS gene was related to the disease, and localized to chromosome 2q31 (Young et al., 1999). The findings that defects in several independent genes result in the same phenotypic outcome indicate that BBS is heterogeneous and that the genes are likely to lie in a common biochemical pathway. Alstrom syndrome (ALMS) shares many features of BBS but without the hand and feet deformities, mental retardation, or hypogonadism of BBS. Alstrom patients also experience deafness and diabetes. The critical region containing a gene or genes for ALMS has been localized to chromosome 2p13-p12, a region proximal to BBS5. The *tubby* mouse shares many features of both BBS and ALMS, including obesity, retinal degeneration, insulin resistance, and hearing loss and has been suggested to serve as an animal of these syndromes. Nevertheless, the *Tub* gene maps to mouse chromosome 7, and the human homologue maps to human chromosome 11p15, outside of the regions previously identified for BBS or ALMS. Functional studies have demonstrated that the *Tub* gene appears to act as a prohomone convertase similar to PC1.

Albright Hereditary Osteodystrophy

Like Prader-Willi syndrome, Albright Hereditary Osteodystrophy (AHO) is an autosomal dominant disease that results from defects in a single imprinted gene. Obesity, short stature, shortened fingers, subcutaneous calcium deposits, mental deficiency, and resistance to parathyroid hormone characterize the syndrome. Endocrine features of AHO are pseudo-hypoparathyroidism, hypogonadism, and hypothyroidsim. Pseudo-hypoparathyroidism type IA (PHP-IA) caused by a heterozygous inactivating mutation in the alpha subunit of the Gs protein and mapped to chromosome 20, contributes to abnormal secretion of the thyroid hormone and gonadotropins (Izraeli et al., 1992). The defect in chromosome 20 is similar to the region in the mouse chromosome 2 where both maternal and paternal imprinting occurs. Mouse studies demonstrated that the *Gnas* gene is imprinted in a tissue-specific manner (Yu et al., 1998). In humans, the maternally inherited mutation results in AHO, hormonal resistance and somatic abnormalities, and the paternally inherited mutation results in pseudopseudohypoparathyroidism (PPHP), normal hormone function yet manifests AHO abnormalities.

Cushing Disease

Cushing disease is characterized by upper body obesity, excess adiposity around the face and neck, and slender arms and legs. Thin, fragile skin that bruises easily and heals poorly is also a major symptom, as are brittle bones, weak muscles, high blood pressure and hyperglycemia leading to type 2 diabetes mellitus. Mental development is normal in Cushing disease but anxiety and depression occur frequently. Women often have excess hair growth and abnormal reproductive function, and men also have impaired sexual function (DeGroot et al., 1995). Cushing syndrome is the result of chronic exposure to excessive levels of cortisol, which is secreted from the adrenals in response to stimulation by adrenocorticotropin (ACTH) released from hypothalamic areas of the brain. Cushing syndrome displays a high level of heterogeneity with both sporadic and familial forms present; thus, it is likely that multiple genes may produce this disease. Overproduction of ACTH occurs primarily due to pituitary tumors, although other types of tumors (small cell lung cancer, thyomas, and pancreatic islet tumors) can produce ACTH as well, and genes involved in tumor formation are likely candidates for Cushing syndrome. Transgenic mice overexpressing leukemia inhibitory factor (*Lif*), a key regulator of the mature hypothalamic-pituitary-adrenal axis, exhibit dwarfism, hypogonadism, and the characteristic fragile skin and truncal obesity of Cushing syndrome (Yano et al., 1998). Chronic secretion of ACTH can also result as a response to generalized glucocorticoid resistance. A 33-year-old man presenting with infertility, hypertension, and hypercortisolism later developed Cushing syndrome and was found to have a heterozygous mutation in the glucocorticoid receptor that reduced receptor binding sites by half and initiated glucocorticoid resistance (Karl et al., 1996). One additional pathway for the development of Cushing syndrome may be through regulation of POMC expression within the brain via the dopamine pathway. Dopamine D2 receptor knockout mice have multiple characteristics of Cushing syndrome, including increased expression of POMC, elevated ACTH levels and adrenal hypertrophy (Saiardi and Borrelli, 1998). Cushing syndrome occurs in dogs, cats, horses, and rodents and with a wide variety of concomitant symptoms present in each species, suggesting that genetic background may mediate the effects of hypercortisolism. Since ACTH is a cleavage product of POMC and a precursor for α-MSH, obesity in this disease may be due to malfunctions in POMC or ACTH processing by PC1, although this pathway has not yet been explored.

Syndrome X

Since the first half of the 20th century, researchers have noted that multiple metabolic abnormalities appear to aggregate in certain individuals. These abnormalities include obesity, insulin resistance, hypertension, dislipidemia, and impaired glucose tolerance/type 2 diabetes. In 1988, Reaven designated this aggregation of chronic metabolic disorders as "Syndrome X," and since that time, the exact definition of this syndrome has been the topic of much controversy (Reaven, 1988). Syndrome X is defined by the World Health Organization (Alberti and Zimmet, 1998) as the combination of type 2 diabetes (fasting -glucose level of 111 mg/dl), impaired glucose tolerance (fasting glucose level of 126 mg/dl), or insulin resistance (lowest quartile of insulin-stimulated glucose uptake or highest quartile of insulin resistance index) with any two of the following: obesity (BMI >30 kg/m^2 and/or waist-to-hip ratio in males >0.90 and females >0.85); hypertension (antihypertensive treatment or blood pressure ≥160/90 mm Hg); dyslipidemia (a fasting triglyceride level of > 150mg/dl or low high-density lipoprotein cholesterol <35mg/dl for men or <39mg/dl for women); or microalbuminuria (overnight excretion rate ≥20 μg/min). Since each of the conditions listed above can and do occur independently of the others, it is unclear whether chronic obesity, for example, leads to the development of insulin resistance, hypertension or hypertriglyceridemia, or whether the reverse is true.

Each of the disorders above has been shown to have a heritable component, and the aggregation of all of these chronic conditions has been hypothesized to be part of an overall "thrifty genotype," in which organisms that are able to maximize fat storage while minimizing utilization of stored energy are favored for survival. While the multigenic nature of Syndrome X in humans is without question and certainly environmental factors such as exercise and nutrition play an essential role in the onset and severity of disease, it is interesting to note that single gene rodent models of obesity, such as Koletsky *f/f* rats, *ob* and *db* mice, and Zucker *fatty* rats, also concurrently manifest diabetes, hyperinsulinemia, dyslipidemia, and/or hypertension. Thus, any single gene lying directly in a pathway regulating fat storage and utilization is a candidate for Syndrome X. The leptin system is particularly likely to be involved in Syndrome X etiology, as leptin infusion has been shown to influence circulating plasma triglycerides and insulin as well as expression of enzymes involved in free fatty acid oxidation (Chen et al., 1996). Leptin also reduces insulin sensitivity and attenuates several insulin-induced activities, including glucose uptake, lipogenesis, and glycogen synthesis (Cohen et al., 1996). How leptin and insulin may modulate or regulate each other *in vivo* is not clear but these rodent models suggest that leptin-insulin interaction may influence both energy metabolism and blood pressure and produce syndromic obesity in humans.

CONCLUSION

This chapter has provided a brief summary of several human obesity syndromes with both single gene and multigenic etiologies. Though the genes and pathways through which the various obesity syndromes develop may differ, several features are common to the syndromes, and these are summarized in Table 2. Because not all syndromes share the same features, it seems clear that obesity can result through many mechanisms. Further study of obesity syndromes in both humans and animals may help to elucidate these mechanisms.

Table 2. Summary of obesity syndromes in humans.

Syndrome	Single or Multiple Genetic basis	Hyperinsulinemia/ Insulin Resistance	Dyslipidemia	Abnormal glucose levels/Type 2 Diabetes	Hypertension	Hypogonadism	Hyperphagia	Mental Retardation	Somatic abnormality	Other
LEP deficiency	S	✓		✓		✓	✓			reproductive abnormalities
LEPR deficiency	S	✓	✓	✓	✓	✓	✓			emotional & social lability
POMC deficiency	S			✓			✓			red hair pigment
PC1 deficiency	S			✓		✓	✓			↓ cortisol ↑ proinsulin & POMC
MC4R deficiency	S					✓	✓			Increased linear growth
Prader-Willi Syndrome	M			✓		✓	✓	✓	✓	abnormal temperature control
Bardet-Biedl Syndrome	M			✓	✓	✓		✓	✓	renal failure; retinitis pigmentosa; polydactyly
Alstrom Syndrom			✓		✓					retinitis pigmentosa; deafness
Albright Hereditary Osteodystrophy	S				✓	✓		✓		osteoporosis seizures
Cushing's Syndrome	M			✓	✓				✓	↑ cortisol ↓ ACTH
Syndrome X	M	✓	✓	✓	✓		✓			cardiovascular disease

Acknowledgments

This research was supported by grant number UR6/CCU617218-01 awarded by the Center for Disease Control and Prevention. Special thanks are given to Cristina Barroso, Daniel Kainer, and Mitzi Laughlin for their help in preparation of the manuscript.

References

Alberti K and Zimmet P (1998) Definition, diagnosis, and classification of diabetes mellitus and its complications. Part 1: Diagnosis, and classification of diabetes mellitus, provisional report of a WHO consultation. *Diabet Med* 15:539-553

Allison DB, Packer-Muner W, Pietrobelli A, Alfonso VC, Faith MS. Obesity and developmental disabilties: Pathoegenisis and treatment. *Journal of Physical and Developmental Disabilities* 1998;10: (3) 215-255.

Barlow D. Gametic imprinting in mammals. *Science* 1995;270:1610-1613.

Brown A, Dolan J, Willi S, et al. Endogenous mutations in human uncoupling protein 3 alter its functional properties. *FEBS Lett.* 1999;464:189-193.

Carmi R, Rokhlina T, Kwitek-Black A, et al. Use of a DNA pooling strategy to identify a human obesity syndrome locus on chromosome 15. *Hum Mol Genet* 1995;4:9-13.

Cassidy S. Prader-Willi syndrome. *Curr Probl Pediatr* 1984;14:1-55.

Cattanach B, Barr J, Beechey C, et al. A candidate model for Angelman syndrome in the mouse. *Mamm Genome* 1997;8:472-478.

Chen G, Koyama K, Yuan X, et al. Disappearance of body fat in normal rats induced by adenovirus-mediated leptin gene therapy. *Proc Natl Acad Sci* 1996;93:14795-14799.

Chua Jr. SC, Chung WK, Wu-Peng S, et al. Phenotypes of mouse diabetes and rat fatty due to mutations in the OB (leptin) receptor. *Science* 1996;271:994-996.
Clement K, Vaisse C, Lahlou N, et al. A mutation in the human leptin receptor gene causes obesity and pituitary dysfunction. *Nature* 1998;392:398-401.
Cohen B, Novick D, Rubinstein M. Modulation of insulin activities by leptin. *Science* 1996;274:1185-188.
Coleman DL. Obese and diabetes: Two mutant genes causing diabetes-obesity syndromes in mice. *Diabetologia* 1978;14:141-148.
Collins S, Kuhn C, Petro A, et al. Role of leptin in fat regulation. *Nature* 1996;380:677.
Considine RV, Sinha MK, Heiman ML, et al. Serum immunoreactive-leptin concentrations in normal-weight and obese humans. *N Engl J Med* 1996;334:292-295.
DeGroot L, Cahill Jr. G, Martini L, Nelson D, eds. *Endocrinology*. Philadelphia, PA: W.B. Saunders Company; 1995.
Dittrich B, Buiting K, Korn B, et al. Imprint switching on human chromosome 15 may involve alternative transcripts of the SNRPN gene. *Nature Genet* 1996;14:163-170.
Farooqi I, Yeo G, Keogh J, et al. Dominant and recessive inheritance of morbid obesity associated with melanocortin 4 receptor deficiency. *J Clin Invest* 2000;106:271-279.
Flier JS. The adipocyte: Storage depot or node on the energy information superhighway? *Cell* 1995;80:15-18.
Frederich RC, Hamann A, Anderson S, et al. Leptin levels reflect body lipid content in mice: Evidence for diet-induced resistance to leptin action. *Nat Med* 1995;1:1311-1314.
Graham M, Shutter J, Sarmiento U, et al. Overexpression of Agrt leads to obesity in transgenic mice. *Nat Genet* 1997;17:273-274.
Gu W, Tu Z, Kleyn PW, Kissebah A, Duprat L, Lee J, Chin W, Maruti S, Deng N, Fisher SL, Franco LS, Burn P, Yagaloff KA, Nathan J, Heymsfield SB, Albu J, Pi-Sunyer FX, Allison DB. Identification and Functional Analysis of Novel Human Melanocortin-4 Receptor Variants. *Diabetes* 1999; 48:635-639.
Gunay-Aygun M, Cassidy S, Nicholls R. Prader-Willi and other syndromes associated with obesity and mental retardation. *Behav Genet* 1997;27:307-324.
Halaas JL, Gajiwala KS, Maffei M, et al. Weight-reducing effects of the plasma protein encoded by the obese gene. *Science* 1995;269:543-546.
Hamilton BS, Paglia D, Kwan AYM, Dietel M. Increased obese mRNA expression in omental fat cells from massively obese humans. *Nat Med* 1995;1:953-956.
Harris RB. Role of set-point theory in regulation of body weight. *FASEB J* 1990;4:3310-3318.
Heo M, Leibel RL, Boyer BB, Chung WK, Koulu M, Karvonen MK, Pesonen U, Rissanen A, Laakso M, Uusitupa MIJ, Chagnon Y, Bouchard C, Donohoue PA, Burns TL, Shuldiner AR, Silver K, Andersen RE, Pedersen O, Echwald S, Sørensen TIA, Behn P, Permutt MA, Jacobs KB, Elston RC, Hoffman DJ, Allison DB. Pooling analysis of genetic data: the association of leptin receptor (LEPR) polymorphisms with variables related to human adiposity. Submitted.
Hervey GR. The effects of lesions in the hypothalamus in parabiotic rats. *J Physiol* 1959;145:336-352.
Hinney A, Schmidt A, Nottebom K, et al. Several mutations in the melanocortin-4 receptor gene including a nonsense and a frameshift mutation associated with dominantly inherited obesity in humans. *J Clin Endocr Metab* 1999;84:1483-1486.
Holder Jr J, Butte N, Zinn A. Profound obesity associated with a balanced translocation that disrupts the SIM1 gene. *Hum Mol Genet* 2000;9:101-108.
Huszar D, Lynch C, Dunmore J, et al. Targeted disruption of the melanocortin-4 receptor results in obesity in mice. *Cell* 1997;88:131-141.
Iida M, Murakami T, Ishida K, et al. Substitution at codon 269 (glutamine to proline) of the leptin receptor (OB-R) cDNA is the only mutation found in the Zucker fatty (fa/fa) rat. *Biochem Biophys Res Comm* 1996;224:597-604.
Ingalls A, Dickie M, Snell G. Obese, a new mutation in the house mouse. *J Hered* 1950;41:317-318.

Izraeli S, Metzker A, Horev G, et al. Albright hereditary osteodystrophy with hypothyroidism, normocalcemia, and normal Gs protein activity: a family presenting with congenital osteoma cutis. *Am J Med Genet* 1992;43:764-767.

Karl M, Lamberts S, Koper J, et al. Cushing's disease preceded by generalized glucocorticoid resistance: clinical consequences of a novel, dominant-negative glucocorticoid receptor mutation. *Proc Assoc Am Physicians* 1996;108:296-307.

Kennedy GC. The role of depot fat in the hypothalamic control of food intake in the rat. *Procedures of the Royal Society of Britain Biological Sciences* 1953;140:578-592.

Krude H, Biebermann H, Luck W, et al. Severe early-onset obesity, adrenal insufficiency and red hair pigmentation caused by POMC mutations in humans. *Nat Genet* 1998;19:155-157.

Krude I, Gruters I. Implications of Proopiomelanocortin (POMC) Mutations in Humans: The POMC Deficiency Syndrome. *Trends Endocrinol Metab* 2000;11:15-22.

Kurtz T, Morris R, Pershadsingh H. Zucker fatty rat as a genetic model of obesity and hypertension. *Hypertension* 1989;13:896-901.

Kuslich C, Kobori J, Mohapatra G, et al. Prader-Willi syndrome is caused by disruption of the SNRPN gene. *Am J Hum Genet* 1999;64:70-76.

Kwitek-Black A, Carmi R, Duyk G, et al. Linkage of Bardet-Biedl syndrome to chromosome 16q and evidence for non-allelic genetic heterogeneity. *Nat Genet* 1993;5:392-396.

Lee G-H, Proenca R, Montez JM, et al. Abnormal splicing of the leptin receptor in diabetic mice. *Nature* 1996;379:632-635.

Leppert M, Baird L, Anderson K, et al. Bardet-Biedl syndrome is linked to DNA markers on chromosome 11q and is genetically heterogeneous. *Nat Genet* 1994;7:108-112.

Lonnqvist F, Arner P, Nordfors L, Schalling M. Overexpression of the obese (ob) gene in adipose tissue of human obese subjects. *Nat Med* 1995;1:950-953.

MacDonald H, Wevrick R. The necdin gene is deleted in Prader-Willi syndrome and is imprinted in human and mouse. *Hum Mol Genet* 1997;6:1873-1878.

Maffei M, Halaas J, Ravussin E, et al. Leptin levels in human and rodent: Measurement of plasma leptin and ob RNA in obese and weight-reduced subjects. *Nat Med* 1995;1:1155-1161.

Miller M, Duhl D, Vrieling H, et al. Cloning of the mouse agouti gene predicts a secreted protein ubiquitously expressed in mice carrying the lethal yellow mutation. *Genes Dev* 1993;7:454-67.

Montague C, Farooql S, Whitehead J, et al. Congenital leptin deficiency is associated with severe early-onset obesity in humans. *Nature* 1997;387:903-908.

Mueller G, Ertl J, Gerl M, Preibisch G. Leptin impairs metabolic actions of insulin in isolated rat adipocytes. *J Biol Chem* 1997;272:10585-10593.

Muoio D, Dohn G, Fiedorek F, et al. Leptin directly alters lipid partioning in skeletal muscle. *Diabetes* 1997;46:1360-1363.

Naggert J, Fricker L, Varlamov O, et al. Hyperproinsulinaemia in obese fat/fat mice associated with a carboxypeptidase E mutation which reduces enzyme activity. *Nat Genet* 1995;10:135-142.

Online Mendelian Inheritance in Man, OMIM (TM). In.: McKusick-Nathans Institute for Genetic Medicine, Johns Hopkins University (Baltimore, MD) National Center for Biotechnology Information and National Library of Medicine, (Bethesda, MD); 2000.

O'Rahilly S, Gray H, Humphreys PJ, et al. Brief report: impaired processing of prohormones associated with abnormalities of glucose homeostasis and adrenal function. *New Eng J Med* 1995;333:1386-1390.

Pellymounter MA, Cullen MJ, Baker MB, et al. Effects of the obese gene product on body weight regulation in ob/ob mice. *Science* 1995;269:540-543.

Phillips M, Liu Q, Hammond H, et al. Leptin receptor missense mutation in the fatty Zucker rat. *Nat Genet* 1996;13:18-19.

Reaven G. Banting lecture 1988: Role of insulin resistance in human disease. *Diabetes* 1988;37:1595-1607.

Roybal R. Obese Girl Suffers From Rare Genetic Defect. In: *Albuquerque Journal.* Albuquerque, NM; 2000.

Saiardi A, Borrelli E. Absence of dopaminergic control on melanotrophs leads to Cushing's-like syndrome in mice. *Mol Endocrinol* 1998;12:1133-1139.

Sheffield V, Carmi R, Kwitek-Black A, et al. Identification of a Bardet-Biedl syndrome locus on chromosome 3 and evaluation of an efficient approach to homozygosity mapping. *Hum Mol Genet* 1994;3:1331-1335.

Shutter J, Graham M, Kinsey A, et al. Hypothalamic expression of ART, a novel gene related to agouti, is up-regulated in obese and diabetic mutant mice. *Genes Dev* 1997;11:593-602.

Strobel A, Issad T, Camoin L, et al. A leptin missense mutation associated with hypogonadism and morbid obesity. *Nat Genet* 1998;18:213-215.

Takaya K, Ogawa Y, Hiraoka J, et al. Nonsense mutation of leptin receptor in the obese spontaneously hypertensive Koletsky rat. *Nat Genet* 1996;14:130-131.

Tartaglia LA, Dembski M, Weng X, et al. Identification and expression cloning of a leptin receptor, OB-R. *Cell* 1995;83:1263-1271.

Vaisse C, Clement K, Durand E, et al. Melanocortin-4 receptor mutations are a frequent and heterogeneous cause of morbid obesity. J Clin Invest. 2000;106:253-262.

Vaisse C, Clement K, Guy-Grand B, Froguel P. A frameshift mutation in human MC4R is associated with a dominant form of obesity. *Nat Genet* 1998;20:113-114.

Williams C, Angelman H, Clayton-Smith J, et al. Angelman syndrome: consensus for diagnostic criteria.Angelman Syndrome Foundation. *Am J Med Genet* 1995;56:237-238.

Wu-Peng X, Chua Jr. S, Okada N, et al. Phenotype of the obese Koletsky (f) rat due to tyr763stop mutation in the extracellular domain of the leptin receptor (Lepr). *Diabetes* 1997;46:513-518.

Yano H, Readhead C, Nakashima M, et al. Pituitary-directed leukemia inhibitory factor transgene causes Cushing's syndrome: neuro-immune-endocrine modulation of pituitary development. *Mol Endocrinol* 1998;12:1708-1720.

Yaswen L, Diehl N, Brennan M, Hochgeschwender U. Obesity in the mouse model of pro-opiomelanocortin deficiency responds to peripheral melanocortin. *Nat Med* 1999;5:1066-1070.

Yeo GSH, Farooqi IS, Aminian S, et al. A frameshift mutation in MC4R associated with dominantly inherited human obesity. *Nat Genet* 1998;20:111-112.

Young T, Penney L, Woods M, et al. A fifth locus for Bardet-Biedl syndrome maps to chromosome 2q31. *Am J Hum Genet* 1999;64:900-904.

Yu S, Yu D, Lee E, et al. Variable and tissue-specific hormone resistance in heterotrimeric Gs protein alpha-subunit (Gs-alpha) knockout mice is due to tissue-specific imprinting of the Gs-alpha gene. *Proc Natl Acad Sci* 1998;95:8715-8720.

Zhang Y, Proenca R, Maffel M, et al. Positional cloning of the mouse obese gene and its human homologue. *Nature* 1994;372:425-432.

Chapter 2

THE SPECTRUM OF EATING DISORDERS IN HUMANS

Janet L. Treasure and David A. Collier
Institute of Psychiatry
South London & Maudsley NHS Trust

ABSTRACT The aim of this chapter is to summarise the clinical features of the eating disorders as they are conceptualised at the beginning of the twenty first century. Anorexia nervosa was recognised as a medical condition in the nineteenth century. The disorders of binge eating emerged during the intervening hundred years. We do not include in the remit of this chapter some of the rarer disorders of eating such as pica or the disorders of childhood (food avoidance emotional disorder, selective eating, functional dysphagia or pervasive refusal syndrome). In this chapter we describe the clinical features of adult eating disorders and focus upon the biological aspects of pathophysiology and aetiology. Thus we summarise the current knowledge into the regulation of appetite, eating and satiation. Also we dip into the latest research into aetiology including genetic vulnerability and environmental stress. Eating is one of the basic primary reinforcers of behaviour and has an impact on pleasure and basal emotional tone. In people with eating disorders, eating or not eating appears to be used as a means of regulating emotion as much as it is used to fulfil its role in nutritional homeostasis. Thus it is interesting to compare and contrast human eating behaviour and its regulation with the control of this behavior in animals.

INTRODUCTION

The aim of this chapter is to introduce the clinical spectrum of eating disorders. This is somewhat of a moving target as the clinical conceptualization has changed rapidly over time. Although there appears to be

J.B. Owen et al. (eds.), Animal Models – Disorders of Eating Behaviour and Body Composition, 19–49.

an innate human tendency to think in terms of categories this may be counterproductive as many of the symptoms and even syndromes are more properly thought about in terms of a spectrum.

What are Eating Disorders?

Most of the other chapters in this book describe the biological underpinnings of eating and the control of nutritional homeostasis. There has been an exponential growth in knowledge within this domain. We are particularly interested in how this increase in knowledge can be applied to disorders of eating in humans. In Western society deviations from the nutritional norm are considered to be a problem of too much, or too little will power. However the biological underpinnings described in this book will make us question this.

In current medical models eating disorders are classified within psychiatry and as such represent an impairment of mental health. However mental health is derived from several domains of life (Figure 1).

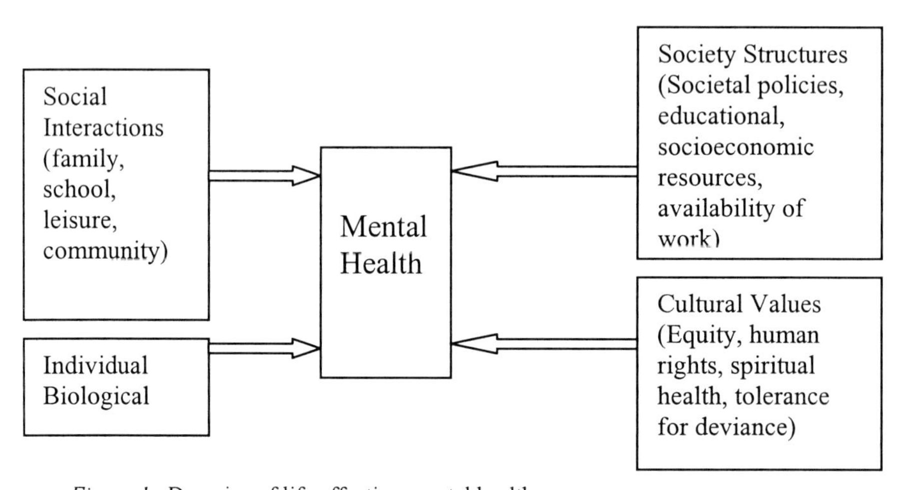

Figure 1 Domains of life affecting mental health

Biological factors contribute to and interact with the various tiers of environmental input, interpersonal, societal and cultural to produce a broad array of unhealthy eating and nutritional balance. It can be difficult to unearth the underlying biological matrix. Thus the issue of mental health is wider than the absence of a mental disorder but includes the concept of psychological resource. This incorporates positive affect, the ability to cope with adversity, the ability to form close and sustained relationships, citizenship, and the sense of belonging in society and connectedness with healthy structures. All of these serve to shape the clinical presentation. In the first part of this chapter we will briefly describe the spectrum of eating disorders, as they are conceptualized at the beginning of the twenty first century. We will focus on the clinical features and what is known about aetiology.

There is uncertainty about the definitions and categorisation of the eating disorders. In contemporary parlance and psychiatric diagnostic systems, for example 'Diagnostic and Statistical Manual of Mental Disorders' (American Psychiatric Association, 1980)) the eating disorders consist of anorexia nervosa and the bulimia disorders.

Table 1a: Criteria for the Classification of Anorexia Nervosa

DSM-IV	ICD 10 (WHO 1992) F50
Refusal to maintain body weight over a minimal norm/leading to body weight 15% below expected	Significant weight loss (BMI<17.5 kg/m^2) or failure of weight gain or growth
	Weight loss self-induced by avoid fattening foods and one or more of the following (a) vomiting (b) purging (c) excessive exercise (d) appetite suppressants (f) diuretics
Intense fear of gaining weight or becoming fat	A dread of fatness as an intrusive overvalued idea and the patient imposes a low weight threshold on herself
Disturbance in the way in which one's body weight, size or shape is experienced eg "feeling fat" {denial of seriousness of underweight or undue influence of body weight and shape on self evaluation}	
Absence of 3 consecutive menstrual cycles	Widespread endocrine disorder (a) amenorrhoea (b) raised growth hormone (c) raised cortisol (d) reduced T3
Restricting type: Binge/purging type: binge eating or vomiting/misuse laxatives diuretics	

The underlying assumption is that psychological /psychiatric factors play a role in the aetiology and /or management. The binge eating disorders entered this diagnostic system in 1990 (American Psychiatric Association, 1994). The criteria for what is now known as bulimia nervosa have evolved and become more specific over time. By default, this led to an increase in the size of the group who do not exactly fulfil these diagnostic criteria, that is, eating disorder not otherwise specified. (Spitzer *et al.*, 1991) suggested that an additional category of binge eating disorder be defined from within this group. This is now in the fourth edition of Diagnostic and Statistical Manual of Mental Disorders (American Psychiatric Association, 1994) as a proposed category pending further research. The diagnostic criteria are shown in Tables 1a and 1b.

Table 1b: Criteria for the Classification of Bulimia Nervosa

DSM-III-R Bulimia Nervosa (307.51)	**ICD-10 Bulimia Nervosa** (F50.1)	**DSM IV Binge Eating Disorder**
Recurrent episodes of binge eating {(1) Large amount of food (2) loss of control}	Episodes of overeating	Recurrent episodes of binge eating (1) Large (2) No control
Feeling of lack of control of eating during binge		Binge eating associated with at least 3 of the following (a) rapid eating (b) uncomfortably full (c) eat without hunger (d) shameful, solitary eating (e) disgust, guilt, depression after overeating
Regular use of methods of weight control a) vomiting b) laxatives c) diuretics d) fasting/strict diet e) vigorous exercise	Methods to counteract weight gain a) vomiting b) laxatives c) fasting d) appetite suppressants e) metabolic stimulants f) diuretics	Marked distress after binge eating
Minimum average of 2 binges a week in 3 months		Binge eating on average at least 2 days a week for 6 months
Persistent overconcern with shape or weight	Morbid fear of fatness with a sharply defined weight threshold	
	Often a history of anorexia nervosa	Disturbance not exclusively present with anorexia nervosa or bulimia nervosa

Italic : { }: proposed addition to DSM IV (Table 1 a&b)

Continuities versus categories?

However when we turn to the clinical symptoms of individual cases the divisions appear to be somewhat arbitrary and based on soft evidence. The borderlines between categories in the overeating ranges are fuzzy. The presenting symptoms are often unreliable. The aetiological justification for producing a split within the spectrum of disorders of eating behaviour is also tenuous in the current state of knowledge. Therefore it may be more appropriate to consider all the eating disorders as lying on a spectrum (Figure 2).

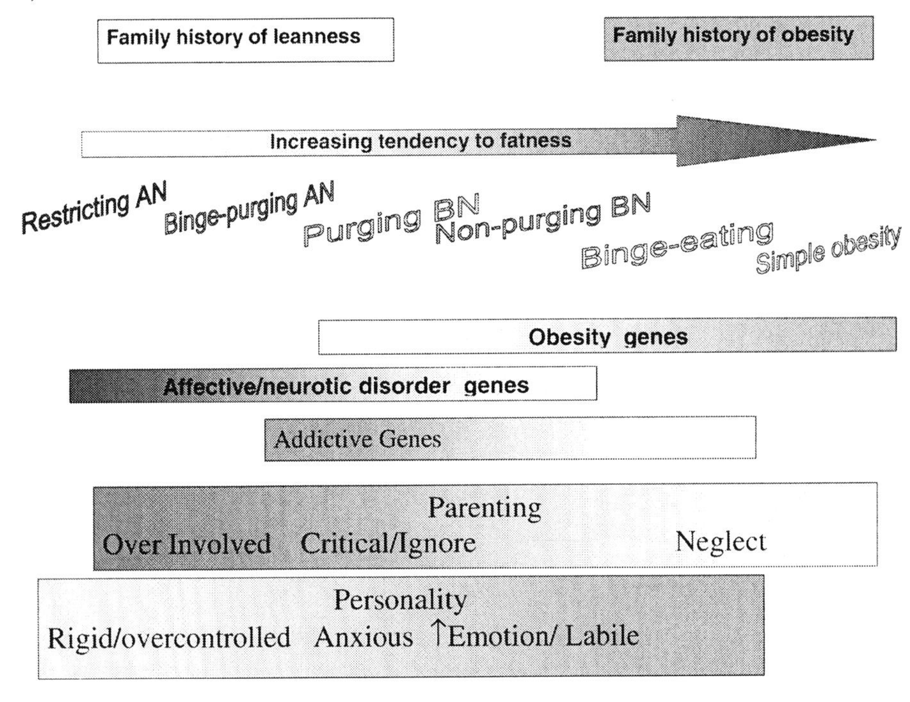

Figure 2 The full range of eating disorders (including obesity)

Others may look to history to explain the present. Obesity was considered to be one of the forms of psychosomatic illnesses suitable for treatment with analytical psychotherapy. Hilde Bruch was an analyst who had a profound effect on our current understanding of anorexia nervosa, having worked from a platform of treating obesity (Bruch, 1973). Obesity has been generally excluded from the remit of psychiatry as belief in analytical therapy waned (However there is a chapter on obesity in the latest edition of the Oxford textbook of psychiatry, Gelder et al., 2000). This contrasts with anorexia nervosa where the converse has occurred. Physicians treated anorexia nervosa, by admission to medical beds, until the later part of the twentieth century. A variety of psychotherapeutic methods, which are effective in the treatment of eating disorders, are now being formulated within the rubric of psychiatry. Thus history and clinical symptomatology can be fickle and fallible and it is perhaps time for a more pragmatic approach. We need to examine our prejudices more carefully. The theoretical purist will want aetiological evidence and the pragmatist will want effectiveness. It is hoped that the knowledge gleaned from basic science, discussed in this book, will stimulate interest and growth in aetiological evidence.

The symptoms or features that determine the borderlines between categories of eating disorders are of three types: the degree of deviation from normal weight (amenorrhoea is a complementary marker for this in anorexia nervosa), the pattern of eating (under to overeating) and the weight control measures that are used (vomiting, laxatives, overexercising etc). There is a degree of asymmetry in the definitions as deviation of weight above the normal range i.e. obesity is not included at the present time.

However emotional factors are a core implicit component of the eating disorders. Eating may serve to regulate emotions in those with a matrix of inter and intra personal difficulties. Thus it has been suggested that the conceptualisation of eating disorder subtypes should be based not only on eating symptoms but also on personality (Westen & Harnden-Fischer, 2001) Indeed, this approach improves the validity of the diagnostic grouping in terms of symptoms, prognosis, adaptive functioning and aetiology. Three types of personality domains emerged in this study which was based on expert ratings traits and parallels the findings in studies which have utilised self report instruments (Goldner *et al.*, 1999). The first was a high functioning self critical, perfectionistic group who tend to feel guilty and anxious and in the main have bulimia, although a few presented with anorexia nervosa. The second cluster was a constricted, overcontrolled profile in which there is a restriction of pleasure, needs, emotions, relationships, self-knowledge, self-reflection, sexuality and depth of understanding of others. These patients are dysphoric, anhedonic, anxious and ashamed. This type of personality is usually linked with people who have restricting anorexia nervosa. The third type of personality cluster includes people who are emotionally dysregulated, undercontrolled and impulsive. They tend to seek relationships to soothe themselves. People with this type of personality have often been sexually abused (however a proportion of those in the constricted cluster has also been abused). The eating disorder symptoms, which are usually in the bulimic domain, tend to be used to modulate intense affect. In childhood these personality subgroups manifest as temperamental traits. For example in the constricted, restricting anorexia nervosa group, childhood traits of rigidity and inflexibility (Brecelj *et al.*, 2001), negative self-evaluation, isolation from peers and perfectionism (Fairburn *et al.*, 1999a) are present. There is less perfectionism but more anxiety in the markers of childhood traits which cluster into the emotionally dysregulated group which go on to develop bulimic disorders (Fairburn *et al*, 1997). This indicates that in part the eating disorder symptoms are a manifestation of temperamental traits and personality dimensions. These in turn are based upon a developmental organisation of affect and cognition used as the template for intra and interpersonal functioning. It is uncertain to what degree animal models will be able to help us understand these aspects of eating disorders in humans.

Epidemiology

As we move to the right on the spectrum of eating disorders in Figure 1 the prevalence within the population changes. The incidence of new cases of anorexia nervosa presenting for any type of medical care is between 4-8/100,000 of the total population (Lucas *et al.*, 1991; Hoek, 1991; Turnbull *et al.*, 1996). The incidence in males is lower than 0.5/100,000. The incidence rate falls off rapidly by a factor of 10 in women over the age of 25. Lucas *et al.*, (1991) estimated the point prevalence of anorexia nervosa to be 0.15% for the total population and 0.48% for 15-19 age year old girls. The incidence of cases of bulimia nervosa presenting to primary care in the UK and Holland is 12/100,000 population and 11/100,000 population (Hoek, 1991;Turnbull, 1996). The incidence of cases presenting to any form of medical care in the USA is 13.5 /100,000 of the total population (Soundy *et al.*, 1995). There is the suggestion that the incidence of bulimia nervosa presenting for medical care has increased during the eighties (Turnbull *et al.,*1996) with an age cohort effect and a marked increase in the risk in those women born after the 1950's (Kendler *et al*, 1991). The prevalence rate of bulimia nervosa in young women has consistently been found to be about 1% in studies which have used the most sophisticated methodology (Whitehouse *et al.*, 1992; Rathner & Messner, 1993). The point prevalence in the total population is 20/100,000 (Hoek, 1991) and the lifetime prevalence was found to be 1% (Garfinkel *et al.*, 1995). The prevalence of Binge Eating Disorder (BED) in the general population ranges between 1-3% (Spitzer *et al.*1991; Spitzer *et al.*, 1993; Fairburn *et al.*, 1998). It is probable the prevalence is higher in the obese population (Geiselman *et al.*, 1999). In the group coming for weight control treatment 1-30% have been reported to fulfil the criteria (Basdevant *et al.*, 1995; Ricca *et al.*, 2000). It is possible that the incidence of BED in certain high risk groups such as individuals with type II diabetes may be as high as 20% (Kenardy *et al.*, 1994)

Clinical Features

As discussed above there is the dichotomy between the clinical diagnostic systems which have introduced each disorder as a new category into the diagnostic fold one at a time and the symptomatic dimensions which appear to lie on a spectrum. If we use the model of the continuum (see Figure 2) we find that some aspects of the clinical features and even etiological risk factors have similar parallel distributions. However there are also discontinuities (Table 2).

For example, the age of onset does not follow a smooth curve. The mean age of onset for anorexia nervosa is 15 and for bulimia nervosa 17 (Turnbull, 1996). People with binge eating disorders have a bimodal age of onset. There is a group who develop their symptoms early at the mean age of 11 and a second group who develop the disorder in their mid twenties (Mussell *et al.*, 1995; Wilfley *et al.*, 1997). Some of the symptoms evolve gradually and are less easy to date than others, for example, the onset of obesity is not always easy to define although it is clear that for a subgroup it emerges in childhood.

It is difficult to get a reliable account of those symptoms, which lie, to the right of the eating disorder spectrum (Figure 1). Truth is lost through the filter of societal values. Our basic understanding of the laws of thermodynamics leads to the argument that obesity results from greed or sloth, two of the seven deadly sins. This stigma makes it difficult for people to admit to problems of overeating. Nutritional diaries are notoriously inaccurate in the obese (although less so in anorexia nervosa). It is uncertain whether this deception is deliberate, simple forgetfulness or more complex unconscious processes. We should give credence to the latter two mechanisms. Research based upon ecological momentary assessments has found that two thirds of those classified by standardized retrospective clinical interviews as "**not** having a form of binge eating disorder" record binges similar in frequency, calorie and nutrient content to the group classified as **having** binge eating disorder (Greeno *et al.,* 2000). This suggests that we may need to add to our list of compensatory behaviors related to binge eating, forgetting. Other compensatory behaviors used to control weight may not be easily elicited as they are considered to be shameful and disgusting.

However despite this difficulty in obtaining evidence of overeating or of weight compensation strategies (and so distinguishing between the categories on the right hand side of the spectrum), there is evidence to suggest that people with binge eating disorder constitute a specific subgroup. Nevertheless many of the differences between bulimia nervosa on one side and obesity on the other may be quantitative rather than qualitative.

Eating Disorder Symptomatology

The most pronounced contrasts between groups at the right of the spectrum are between binge eating disorder and obesity. It may be somewhat of a tautology to note that people with binge eating disorder have more eating disorder psychopathology than people with obesity (Wilfley *et al.*, 2000). Such people have an earlier onset of obesity and weigh more (Telch & Agras, 1994; Yanovski *et al.*, 1993). The levels of psychopathology observed in individuals with BED fall between the high levels seen in Bulimia nervosa (BN) and the low levels seen in simple obesity (Fichter *et al.*,1993). Emotional and personality difficulties are common and there is often poorer work and social functioning. (Fichter *et al.*,1993; Telch & Stice, 1998)

APPETITE, HUNGER & EATING

Many experts dismiss the subjective and behavioral measures that are used as markers of appetite in people with eating disorders as unreliable and distorted. For example, Palmer suggests that appetite in anorexia nervosa is normal but kept under tight control (Palmer, 2000). Others suggest that there must be some abnormality in the appetite control system in order to sustain such a negative homeostatic balance (Pinel *et al.*,2000). It is somewhat surprising that hunger and appetite in eating disordered people has received little systematic study so that each of these divergent views can be held with impunity.

Hunger, Appetite & the regulation of Eating in Anorexia Nervosa

People with anorexia nervosa have higher levels of subjective fullness than a comparison group both before and after a meal (Robinson, 1989). Furthermore people with anorexia nervosa display all the markers which go with decreased hunger and increased satiety. For example, the rate of eating is slowed and there are more pauses within the meal so that meals are prolonged (Sunday & Halmi, 1996). They salivate less than controls in response to olfactory food stimuli (LeGoff *et al.*,1988). The subjective perception of fullness in anorexia correlates with gastric content as would be expected (Robinson, 1989) but gastric emptying is delayed (Robinson, 1988).

Pinel and colleagues suggest that reduced eating in anorexia nervosa may be caused by a decline in the positive incentive value of eating food. Indeed, the reward value of tastes is lower in people with anorexia nervosa (Drewnowski *et al.*, 1987; Sunday & Halmi, 1990) and preference for foods with a higher fat content is lower (Simon *et al.*, 1993). However there is not a blanket, non-specific, rejection of food. Food preferences vary appropriately with the metabolic context. This suggests that the response to food is subject to biological regulation albeit around an abnormal set point.

Not only do foods have less reward value but also the pleasurable emotional reaction to food is reversed to fear and disgust. People with anorexia nervosa report anxiety when shown images of food (Ellison *et al.*, 1998). Electrodermal activity increases during a meal (Leonard *et al.*, 1998), which contrasts with the fall in electrodermal activity that occurs when a comparison group ate a meal in similar circumstances. Also an increase in autonomic arousal (as indicated by increased skin conductance) occurs during the Stroop "food task" (Leonard *et al.,*1998). The latter is completed more slowly by people with anorexia nervosa which indicates that food words interfere with cognitive tasks (Channon, 1988).

In summary, the reaction to food in anorexia nervosa is unusual, as there is 1) decreased preferences 2) a negative subjective emotional response 3) autonomic arousal and 4) attentional bias.

Neuroendocrine response

Various abnormalities in the neuroendocrine response to a meal have been reported in anorexia nervosa. Insulin or glucose response to a meal is reduced in the acute state (Uhe *et al.*, 1992; Tamai *et al.*, 1991) and after recovery (Brown *et al.*, 2001). Pancreatic polypeptide levels (Uhe *et al.,*1992) and growth hormone (GH) are increased (Tamai *et al.,*1991). Research reports into CCK release have been inconsistent. Some groups find a normal response (Pirke *et al.*, 1994) (Geracioti *et al.*, 1992) whereas others report increased CCK (Tamai *et al.*, 1993; Harty *et al.*, 1991; Phillipp *et al.*, 1991). It is uncertain how to account for these discrepancies. Possible reasons include the specificity of the assay, the type of meal or patient!

Leptin levels are decreased in anorexia nervosa (Baranowska *et al.*, 2000; Monteleone *et al.*, 2000a; Lear *et al.,*1999; Nakai *et al.*, 1999) and the levels of soluble leptin receptor are increased (Jiskra *et al.*, 2000). The binge purge subtype of anorexia nervosa tends to run higher levels of leptin than the restricting subtype but this is not surprising as their weight is usually higher (Mehler *et al.,*1999). Leptin levels correlate with fat cell mass and body mass index as would be expected. Leptin levels increase in anorexia nervosa with weight gain (Polito *et al.*, 2000). There is a very weak tendency for leptin levels to be lower after full recovery in anorexia nervosa but this may be related to the lower fat mass that such patients tend to have (Frey *et al.*, 2000; Epel *et al.*, 2001). No differences were found in CSF levels of leptin, neuropeptide Y and peptide YY after full recovery from all types of eating behaviour (Gendall *et al.*, 1999)

Hunger and Appetite in Bulimia Nervosa

The pattern of eating in people with bulimia nervosa is variable and dependent upon the environmental context. Under laboratory conditions people with bulimia nervosa restrict their intake (Rolls et al., 1997} However if their eating is monitored over several days in an environment when they are free to binge and vomit then they eat significantly more than a comparison group (Hetherington *et al.*, 1994). In contrast to anorexia nervosa the positive incentive of food is increased. People with bulimia nervosa rate sweet tastes as more pleasant than a comparison group (Franko *et al..*,1994) and salivate more to olfactory stimuli than did a comparison group (LeGoff *et al.,* 1988).

Cues for eating

Restraint theory was one of the most influential models explaining the mechanisms of overeating in bulimia nervosa (Polivy & Herman., 1985). This theory underpinned the cognitive behavioural model used in treatment (Wilson & Fairburn., 1993). The essence of this model was that dietary restraint, during which weight fell below the biological set point or in the shorter term when the hunger drive was not inhibited by signals of satiety, predisposed to overeating. This model is much less compelling in the case of binge eating disorder when body weight is usually above the presumed biological set point. Furthermore, in forty percent of cases there is no evidence that dieting or weight control precedes the onset of the disorder. Some people recall overeating when they were as young as 2 years old. The assumption of a homeostatic set point control of body weight would be that the hunger drive and predisposition to eat would be attenuated in states of overnutrition. Thus something other than forces regulating weigh drives the overeating.

Food is one of the primary innate reinforcers. The detection of an internal imbalance such as low blood sugar in the case of hunger and the achievement or even the prospect of rectifying this imbalance usually initiates pleasure.

Thus eating and related behaviors are tightly linked to positive emotions. It is therefore not surprising that emotional factors rather than hunger can trigger eating behaviors in some circumstances (reviewed by Jarman & Walsh, 1999). People with bulimia nervosa respond to stressful imagery (personal rejection and loneliness) with an increase in hunger and the desire to binge (Tuschen-Caffier & Vogele, 1999).

Satiety

Meals (or glucose) produce less satiety in people with bulimia nervosa (Kissileff *et al.*, 1996) A counter regulatory response occurs. The greater the size of the meal the stronger the desire to binge eat and a negative emotional reaction follows (Blouin *et al.*, 1993; Devlin *et al.*, 1997). Both the perception of gastric fullness and the rate of gastric emptying are abnormal in bulimia nervosa (Koch *et al.*, 1998). People with bulimia nervosa have a blunting of cholecystokinin release following a meal (Geracioti & Liddle., 1988). If purging occurs insulin and glucose fall rapidly, thus maybe serving to perpetuate the pattern (Johnson *et al.*, 1994).

Reduced leptin has been found in bulimia nervosa (Brewerton *et al.*, 2000; Monteleone *et al.*, 2000; Jimerson *et al.*, 2000) even when weight and BMI were controlled for, although some studies have found normal levels (Ferron *et al.*, 1997). Brewerton *et al.* (2000) found that leptin levels were negatively correlated with baseline cortisol but this result was not replicated by Monteleone *et al.* (2000) and was positively related to prolactin release following 5HT challenges. The reduction in leptin levels produced by acute fasting was reduced in people with bulimia nervosa (Monteleone *et al.,* 2000). Low leptin levels persist after recovery from bulimia nervosa (Jimerson *et al.,*2000)).

In conclusion the reaction to food in bulimia nervosa contrasts markedly to that in anorexia nervosa in that food retains its positive incentive but the reaction to food is dysregulated in that the normal satiety response does not occur.

Binge Eating Disorder (BED)

In a food laboratory people with binge eating disorder BED eat more, in particular of dessert and snack food, than people with simple obesity (Goldfein *et al.*, 1993) but eat smaller quantities than those with bulimia nervosa (LaChaussee *et al.*, 1992). People with binge eating disorder have more difficulty in interpreting visceral sensations related to hunger and satiety (Eldredge & Agras., 1997)

Leptin levels are increased in binge eating disorder (Monteleone *et al.*, 2000b)

The Neural Correlates of Impaired Appetite: Brain Scanning over the eating disorder spectrum

The central information processing pathways for food stimuli in the visual, olfactory and gustatory modalities in the primate brain have been studied in depth. Rolls & Baylis (1994) have summarised the findings from their meticulous series of experiments involving electrophysiological recording from individual neurones. The initial processing of food stimuli occurs in the inferior temporal visual cortex (visual), olfactory bulb and piriform cortex (smell), and nucleus of the solitary tract, thalamus and insula (taste). These areas project to the amygdala, orbito frontal cortex, lateral hypothalamus and striatum where the reward value of the stimuli is calibrated. The reward value assigned to food related stimuli are modulated by integrated information from peripheral metabolism and the gastrointestinal tract.

Dopaminergic projections to the ventral striatum and other limbic structures play a major role in food reward and food driven behaviour (Wise & Rompre,1989). The lateral hypothalamus, in particular the perifornical area, is also implicated in the modulation of the rewarding aspect of food. For example, weight loss (25% body weight) enhances the rewarding effect of stimulation in this area whereas leptin administration produces the opposite effect and attenuates the effectiveness of the rewarding stimulus for up to four days (Fulton *et al.,* 2000). Thus this particular reward system is only responsive to long term shifts in energy balance. Acute food deprivation for up to 48 hours does not have an effect on these neurones. This suggests that in states of chronic starvation, with low leptin such as anorexia nervosa, areas of the brain associated with food reward should become more sensitive to food stimuli.

Brain imaging of the appetite control system in man.

Hunger (provoked by a 36 hour fast) is associated with increased activity in the hypothalamus, insular cortex, limbic and paralimbic areas (medial frontal lobe cortex, anterior cingulate, hippocampus and parahippocampus) during PET scanning (Tataranni *et al.*, 1999). Food versus non-food stimuli produced decreased flow in left temperoinsular cortical regions (Gordon *et al.*, 2000). Thus both external or internal appetite cues are associated with changes in blood flow in the insular cortex especially on the left.

In the satiated state following a liquid meal (containing 50% of the resting energy expenditure) flow in the prefrontal cortex and the inferior parietal lobule increased (Gautier *et al.*, 1999). Flow in the insular cortex and the orbitofrontal cortex decreased and correlated with insulin levels after the meal. Zhao *et al.* (2000) found the decrease in hypothalamic activity measured by fMRI, ten minutes after the consumption of 75 g of glucose also correlated with insulin but in this case, at the fasting level. Thus the metabolic flux, in particular changes in insulin, associated with a meal reduces cerebral blood flow in the hypothalamus, insula and orbitofrontal cortex.

Scanning in Anorexia Nervosa

A variety of functional scanning studies have examined the reaction to food in anorexia nervosa. People with anorexia nervosa had higher relative activity in the left hemisphere, particularly in the inferior frontal and left temporal regions, when viewing cake. This contrasts to the right-sided activation in controls (Nozoe *et al.*, 1993). A more anatomically detailed study using functional magnetic imaging found differential activation in anorexia nervosa between the high/ low calorie drink contrasts in the left limbic system (amygdala, hippocampus, paralimbic system, insula and prefrontal cortex) (Ellison *et al.,*1998). In a second study from their laboratory, food vs non-food stimuli produced differential activation in the left orbitofrontal cortex. Drink versus non-drink stimuli produced increased activation in both the orbitofrontal cortex and insula. Thus the visual presentation of food in people with anorexia nervosa elicits a specific neural response in the left insula and frontotemporal areas. It is uncertain whether this increased flow represents increased hunger, the expected homeostatic response in people who are underweight, or whether it follows from the reversed emotional response to food.

Scanning in Bulimia Nervosa

Nozoe *et al.*.(1995) in their SPECT study of 5 women with bulimia nervosa found increased flow in the left temporal and right inferior frontal regions before feeding. After feeding there was reduced flow bilaterally in the inferior frontal, and in the left temporal and right parietal areas.

Scanning in Binge Eating Disorders

Karhunen *et al.* (2000) also using SPECT found that exposure to food was associated with different changes in the cerebral blood flow in obese women who binge ate compared to either normal weight or obese controls. The binge-eating group had increased flow in the left hemisphere especially in the frontal and pre-frontal regions. There were strong correlations between frontal flow and hunger during exposure to the food. In a study following a person with binge eating disorder in the binge eating state there was a bilateral increase in global blood flow whereas in the restrictive phase flow on the right hand side of the brain was diminished relative to that on the left (Hirano *et al.*, 1999)

Scanning in Obesity

The response (as measured by PET) to a standard meal after a 36 hour fast between obese and lean men (previously reported above) was compared (Gautier *et al..*, 2000). Satiation was associated with an increase in blood flow in the ventromedial and dorsolateral prefrontal cortex and a decrease in the limbic/paralimbic areas and these changes were more pronounced in the obese men than in lean. On the other hand the decrease in flow in hypothalamus and thalamus was attenuated in the obese group. This suggests that brain response to satiation differs in the obese.

Neuroimaging techniques have been used to examine the neuropharmacology of eating disorders. An inverse relationship was found between BMI and striatal dopamine receptor availability (Wang *et al.*, 2001).

Conclusion from Neuroimaging

Thus across the spectrum of eating disorders there are changes in flow in the frontal and limbic areas (where the incentive value of stimuli are processed) in response to either extrinsic or intrinsic food related cues. This line of research is still in its early stages. However it promises to be a useful tool to examine appetite in man.

AETIOLOGY

Let us turn from the immediate antecedents and consequences of the spectrum of eating disorders to predisposing factors. Both extrinsic risk factors and intrinsic genetic vulnerability have been implicated in all conditions.

Anorexia Nervosa

Genetic Risk Factors

It is only recently that the possibility of a genetic predisposition to eating disorders has been a serious subject of research interest. Many of the studies in this area are small or are limited because eating disorders have been added as a supplementary facet to a study of another condition.

Twin Studies

The reliability of twin studies is limited because of low power. The best fit model for a broad anorexia-like phenotype (n=77 out of 1030 twin pairs aged 29) from the Virginia twin registry gave a heritability of 58% (95% CI 33-84%) (Wade *et al.*, 2000a). In a second cohort of 672 female twins (age 17) (also from the USA) the broad anorexia like phenotype (n=26) had a heritability of 74% (95% CI 33-94%) and non-shared variance of 26% (95% C.I. 6-67%) (Klump *et al.,* 2000). In yet a third cohort of a population of 34,142 young Danish twins with self reported anorexia nervosa or bulimia nervosa (total group n=1270) the heritability of narrowly defined AN was 0.48, broad AN was 0.52 and bulimia nervosa was 0.61 (Kortegaard *et al.*, 2001). When the broader phenotype of disordered eating attitudes and behaviours and BMI was modelled in two different age cohorts of twins (age 11 and 17) the pattern of variance differed. In the younger population shared environment dominated over genetic effects but this was reversed in the older cohort (Klump *et al.,* 2000). The majority of the genetic variance on eating attitudes and behaviours in the older group was independent of BMI. Thus it is possible that a genetic risk factor which increases the risk of abnormal attitudes to eating is switched on by the hormonal changes at puberty.

Some of the genetic risk for eating disorders may be shared with other forms of psychopathology. For example there was a shared variance between anorexia nervosa and major depression of 34% (Wade *et al.*, 2000b). Unaffected MZ cotwins (n=19) of people with anorexia-like phenotype had a increased risk of anxiety disorder compared to control MZ twins (n=394) (OR 2.1 (χ^2 = 4.5 p=0.003) (Klump *et al.*, 2001).

Unfortunately many of the family studies have not investigated the body size of family members. Hebebrand & Remschmidt (1995) marshal the evidence that there is a predisposition to leanness in the families of people with anorexia nervosa. It is not thought that this leanness is due to subclinical anorexia nervosa although the risk of anorexia nervosa (adjusted hazard ratio 11.4 95% CI 1.1-89 $\chi^2 = 5.3$ p=0.03) and bulimia nervosa (adjusted hazard ratio 3.5 95% CI 1.1-14 $\chi^2 = 3.9$ p=0.05) is increased in first degree relatives of people with anorexia nervosa (Strober *et al.*, 2000)

People with anorexia nervosa (n=93, age 24 years) have first degree relatives with a higher adjusted risk ratio of major depressive disorder (2.3 95% CI 1.1-4.8), generalised anxiety disorder (3.1 95% CI 1.5- 6.8), obsessive compulsive disorder (4.1 (1.4-12.2) and obsessive compulsive personality disorder (OCPD) (3.6 95% 1.6-8) (Lilenfeld & Kaye., 1998). The rate of OCPD was similar in the relatives whether or not the person with anorexia nervosa was comorbid for OCPD. This suggests linked transmission for OCPD but not for the other forms of psychopathology.

Candidate Genes: Association Studies

An association has been found with the allele polymorphism AN (-1348G/A) in the promoter of the 5-HT2A gene (Collier *et al.*, 1997). However a meta-analysis of all published data showed no overall statistically significant association with AN and the 5-HT2A gene (Odds ratio of 1.26 (1.00-1.90)) (Ziegler & Gorg., 1999). The most obvious explanation for this is that there is no association between anorexia nervosa and the 5-HT2A gene. Failure to replicate might be accounted for by variations in the phenotype between studies; strongest association was seen with the restricting subtype of AN (Enoch *et al.*, 1998).

Overall there has been a failure to find an association with other sites in the 5HT system for example 5-HTT, 5-HT7, tryptophan hydroxylase receptor (TPH) in anorexia nervosa (Campbell *et al.*, 1998; Han, *et al.*, 1999; Hinney *et al.*,1997; Hebebrand *et al.*, 1999). Negative results have been found for the D3 (Bruins-Slot *et al.*., 1998) and the D4 receptor genes (Hinney *et al.*, 1999). Negative results have also been found in relation to genes involved in aspects of nutritional balance: neuropeptide Y5 and Y1 (Rosenkranz *et al.*., 1998a), the β3 adrenergic receptor gene (Hinney *et al.*, 1997), the melanocortin-4 receptor gene (Hinney *et al.*, 1999)and the leptin gene (Hinney *et al.*, 1998).

There have been some positive findings in the D11S911 allele near the UCP-2/UCP-3 gene (Campbell *et al.*, 1999)and the oestrogen receptor β 1082/G allele(Rosenkranz *et al.*, 1998b) across the extremes of weight.

It is impossible to draw conclusions from these studies at the present time. Many of these studies are small and the case mix has been heterogeneous. Larger will be needed to replicate these findings with a clear separation of the various phenotypes.

Early environmental risk factors

A case linkage study from Sweden found an increased risk of anorexia nervosa (ascertained from a psychiatric inpatient register) (n= 781) in girls born with a cephalohematoma (OR 2.4 95% CI 1.4-4.1) and those with prematurity (OR 3.2 95% CI 1.6-6.2) especially if the baby was small for gestational age (OR 5.7 95% CI 1.1-28.7) (Cnattingius *et al.*, 1999). The parents of people with anorexia nervosa show high concern parenting (χ^2 = 17.2 p=0.002) (Shoebridge & Gowers., 2000). In a case-control study involving retrospective reporting of childhood risk factors, the personal vulnerability domain (χ^2 = 52.2 p=0.001) and the environmental domain (χ^2 = 17.5 p=0.001) contributed to the model of risk (Fairburn *et al.*, 1999b). The "dieting domain" had no independent effect. This differentiated the risk model for anorexia nervosa from that found for bulimia nervosa. A similar model was derived when non-affected sisters (n=45) were used as the comparison group (Karwautz *et al.*, 2001).

Chronic stress later in development appears to precipitate the development of anorexia nervosa. For example severe difficulties (16/72 (22%) a vs. 1/28 (4%)) (Schmidt *et al.*, 1997) and adversities (North & Gowers., 1999)precede the onset of anorexia nervosa.

In conclusion genetic factors may predispose to the development of anorexia nervosa possibly in whole or in part by contributing to the childhood temperamental traits which are noted to be important risk factors. However environmental events during development such as stress at birth and traumas such as childhood abuse or difficulties prior to the onset of the illness also appear to contribute as does the nurturing environment. For example overinvolved, overanxious parenting may not be able to buffer the effects of these environmental stressors.

Risk Factors for Bulimia Nervosa

Genetics

A review of studies of bulimia nervosa suggests that between 31-81% of the variance is accounted for by additive genetic effects (Bulik *et al.*, 2000). There is thought to be a common genetic variance of approximately 20% between depression and bulimia nervosa (Walters & Kendler., 1995)

The rate of parental obesity in women with bulimia nervosa or binge eating disorders is increased (OR 4) (Fairburn *et al.,* 1997). The risk of anorexia nervosa (adjusted hazard ratio, 12.1 95% CI 1.5-97) is increased in the first-degree relatives of people with bulimia nervosa. (n=171) suggesting that there common genetic vulnerability to both ends of the eating disorder spectrum (Strober *et al.,* 2000). Lilenfeld & Kaye (1998) review the evidence from family studies: the relative risk of mood disorders is increased two fold in the studies and the relative risk of substance abuse is weakly increased between one to two fold.

Early Environment: Childhood Risk Factors

People with bulimia nervosa have often been exposed to parental depression, alcohol, substance abuse during childhood (OR 9-10) (Fairburn *et al.,*1997). They also report that these patients had low parental contact and yet high parental expectations, in particular about appearance. They also experience critical comments by the family about weight, shape and eating. Abuse from adults in the form of sexual and physical abuse is more common in the bulimic disorders but this is not a strong effect (Fairburn *et al.,* 1998). The major difference in the risk factor model between anorexia nervosa and the binge eating disorders is in the dieting domain (Fairburn et al., 1999a).

Binge Eating Disorder

There has been less research into the aetiology of binge eating disorder. The early environmental risk factors for binge eating disorder are similar to those for bulimia nervosa but abuse from peers in the form of bullying was more common with those with binge eating disorders (Fairburn *et al.,*1998)

In conclusion genetic factors common to both obesity and anorexia nervosa may increase the risk of developing a bulimic disorder. It is possible that the genetic risk shared with anorexia nervosa is that which predisposes to high anxiety and low self-esteem. Perfectionism may be compensatory safety behaviour. Adversity during childhood and chronic stress increases the risk (Schmidt *et al.,*1997). Their childhood experience may be marred by parental psychiatric illness, alcohol, depression etc which may contribute to the neglect or conditional parenting that they experience. There is often criticism about their appearance or behaviour.

Obesity

The genetic risk factors for obesity have been reviewed elsewhere in this book. In a prospective study which examined the effect of risk factors on the development of obesity it was found that parental neglect greatly increased the risk in comparison with harmonious support (Odds ratio 7.1 (95% CI 2.6-19.3). Dirty and neglected children had a much greater risk of adult obesity than aversely groomed children (9.8 (95% CI (3.5-28.2). The risk of obesity was not affected by family structure, over protective parenting or being well-groomed (Lissau & Sorensen, 1994)

Conclusion

Interestingly, in parallel with the spectrum of eating behaviours there appears to be a spectrum of parenting behaviours. High concern, over anxious parenting during the perinatal period is a risk factor for anorexia nervosa. (Shoebridge & Gowers, 2000). This is experienced as an over-involved style of parenting (Fairburn *et al.,*1999a; Karwautz *et al.,* 2001). Conditional parenting characterised by a judgmental style with extremes of neglect and over-involvement is characteristic of bulimia nervosa. In contrast, parental neglect is linked to the obesity (Lissau & Sorensen, 1994). It is possible that it is the pattern of parenting, which perhaps in turn may be shaped to some extent by the temperament of the child, that leads to some of the specificity of eating disorder behavior, as the environmental risk factors do not differ between the groups.

NEUROENDOCRINOLOGY AND NEUROTRANSMITTERS

The Stress System: The Hypothalamic Pituitary Adrenal Axis

Abnormalities in the regulation of the pituitary adrenal system have been implicated across the spectrum of eating disorder.

Our group has developed a neurodevelopmental model of anorexia nervosa. The essence of the model is that stress early in life such as birth injury, in a parental environment which provides a poor buffer for stress for example high levels of parental anxiety leads to a dysregulated and oversensitive stress system (Connan *et al.*, 2000). Stress later in life results in high levels of CRH release, which is unresponsive to cortisol feedback. A dysregulated stress response is not specific for anorexia nervosa but is also seen in depression (McQuade & Young, 2000). The specificity in this model is that in anorexia nervosa there is no secondary recruitment of AVP and upregulation of AVP receptors rather the system continues to be driven by CRH alone. CRH produces a persistent inhibition of appetite whilst the AVP/CRH chronic stress response allows nutritional homeostasis to re-establish some degree of nutritional homeostasis perhaps at a lower level. The behavioural, emotional and physiological effects of CRH bear a striking resemblance to the clinical presentation of AN, including reduced feeding and weight, reduced sexual behaviour; cardiovascular changes; increased anxiety behaviour and reduced social interactions (Dunn & Berridge, 1990). It is not certain what causes the specific lack of AVP recruitment. One possibility is that this results from a genetic vulnerability and the other is that the timing of the effect during the pubertal period is critical. A hormonal trigger could exacerbate the risk. However it is unlikely that puberty alone is sufficient to account for the specificity as the transition to Tanner Stage III at puberty is also associated with an increased incidence of depression in females (Angold *et al.,* 1998).

We have some data, which supports this hypothesis. For example, we have weak evidence that a heightened HPA axis response to stress persists after recovery from anorexia nervosa (Connan *et al..,* 2000; Watson *et al..*, 1999). Also, importantly, in contrast to depression, there is a negative response to the combined DXM/CRH challenge test. This test measures the potentiation of pituitary ACTH release by secretagogues such as vasopressin. This test remains abnormal after recovery from depression (Zobel *et al.*, 2001) and in high-risk relatives (Modell *et al.*, 1998). and is thought to be a marker of trait vulnerability to depression. This differentiates the trait vulnerability in the HPA axis of anorexia nervosa from that of depression. Also we have found that the cortisol response to exogenous AVP is blunted in acute AN, and remains weakly abnormal after full recovery (Connan *et al.,* 2000). This supports the hypothesis that the recruitment of AVP into the stress response does not occur in anorexia nervosa and may explain the close links between anorexia nervosa and depression in terms of the shared genetic vulnerability (Wade *et al.,* 2000b)and childhood vulnerability (Fairburn *et al.,*1999) in that a dysregulated stress response may be a necessary mediating factor.

Neurotransmitters

A large variety of neurotransmitters are involved in the control of appetite. Most of the work in humans has of necessity used indirect methods to measure or modulate neurotransmitter activity. Such studies are often difficult to interpret. The use of scanning in neuropharmacological work may make this area of research in man easier. A serendipitous finding which may provide a clue to the neurochemical underpinning of appetite in man is the finding that the newer antipsychotic drugs, which have less affinity for the D2 receptors than the traditional drugs, and also are antagonists at a variety of other receptors, in particular several 5HT sites, are associated with a rapid increase in weight. Most work on neurotransmitter function in eating disorders has focused on 5HT

Anorexia nervosa

Several studies have examined the role of the central 5HT, which has a key role in appetite, amongst other things, in anorexia nervosa. There have been reports that in anorexia nervosa plasma tryptophan levels are reduced (Coppen *et al.*, 1976) as are CSF levels of 5HIAA (Kaye *et al.*, 1988). The release of prolactin is blunted to various 5HT challenges in acute anorexia nervosa (Goodwin *et al.,* 1989; Brewerton & Jimerson., 1996). After recovery prolactin release to 5HT probes are normalised (O'Dwyer *et al.,*,1996) in the acute response although Crisp and colleagues have some evidence that the response at 5 hours post the test is abnormal (Ward *et al.*, 1998).

Also in acute (Brewerton & Jimerson, 1996) and long term recovery (Ward *et al.,* 1998) meal size does not change in response to 5HT probes. This suggests that some aspects of 5HT function may be dysregulated in anorexia nervosa.

Bulimia nervosa

There has been very little work investigating the role of the HPA axis in bulimia nervosa. Some studies report pathophysiological changes similar to those found in anorexia nervosa (Goldbloom *et al.*, 1989) whereas others report normal values (Walsh *et al.*, 1987; Fichter *et al.*, 1990).

One hypothesis for the development of bulimia nervosa is that a low turnover of 5HT is linked to bulimia nervosa. Dieting reduces tryptophan (TRP) availability, which in turn reduces central 5HT function and may trigger the development of bulimia nervosa in susceptible individuals (Cowen & Smith, 1999). Acute tryptophan depletion led to an increase in calorie intake and irritability in bulimia nervosa (Bastiani *et al.*, 1995) and decreased mood, increased rating in body image concern and subjective loss of control of eating in people who had recovered from bulimia nervosa (Smith *et al.,*1999).

The prolactin response to 5HT probes is reduced in bulimia nervosa (Jimerson *et al.*, 1992; Katzman *et al.*, 1996).

HPA axis in Obesity

Bjorntorp and colleagues have argued that abnormalities in the HPA axis may underpin the development of obesity, especially that associated with visceral obesity, syndrome X (Bjorntorp & Rosmond, 1999). They describe a pattern of low morning cortisol concentrations, an absence of a circadian rhythm, a relative resistance to dexamethasone inhibition and a poor lunch-induced cortisol response. This is thought to be a characteristic consequence of frequently repeated or chronic environmental stress patterns (for example, this pattern was seen in people of a lower socioeconomic class) which is associated with visceral obesity (Rosmond & Bjorntorp, 2000). A similar link between stress reactivity and abdominal obesity has been found in women. Lean women with a high waist to hip ratio reported more chronic stress and in the laboratory found stressful tasks more threatening, released more cortisol to these tasks and did not habituate when they were repeated (Epel *et al.*, 2000). It is uncertain whether the abnormal regulation of the HPA axis in these individuals is linked to depression. However there is obviously a degree of overlap between these phenotypes.

Genetic factors may increase the risk of developing this particular pattern. For example, abnormalities in the glucocorticoid receptor gene with a polymorphic restriction site at the GR gene locus leads to poor feedback regulation of the HPA axis (Rosmond & Bjorntorp, 2000; Bjorntorp & Rosmond, 2000a). Bjorntorp & Rosmond (2000b) suggest that antidepressant drugs, which suppress HPA activity, may have potential to decrease abdominal fat mass.

This theory needs to be evaluated but it is probable that many people with the bulimic disorders would also show this pattern of stress response.

CONCLUSION

It is unclear where the borderlines of eating disorders in humans now lie. For example it could be argued that the eating disorders and obesity could be conceptualised as lying on a continuum. It is probable that there is a complex interplay between genes and environment in all disorders of eating and body composition. There will be both quantitative and qualitative differences between individuals. For example, some people will have a higher loading on the genetic risk factors and yet others may encounter more severe environmental risks or will be in an environment that either exacerbates or amelriates their effect. Nevertheless it is interesting to note that some of the risk factors lie on a continuum. Moreover for some variables there appear to be dose response relationships. For example, the level of parental involvement relates to eating behaviour, in that anxious overinvolvement predisposes to anorexia nervosa whereas neglect predisposes to overeating. Lean genes appear to run in families with anorexia nervosa whereas fat genes run in families with overeating. Other risk factors do not show these dose related effects but may produce specificity by virtue of their temporal impact. For example stress in the perinatal period may predispose to the under eating of anorexia nervosa whereas stress later in childhood may predispose to bulimia nervosa. Mental health professionals have skills in helping people process and cope with the environmental adversities they encounter and are fitted to help that subgroup of people with these risk factors. Whereas professionals more skilled in promoting lifestyle change may be more fitted to intervene with those with a higher loading on the genetic risk factors. The optimal solution may be to have multidisciplinary clinics in which it is possible to match the individuals needs to therapeutic skills.

References

American Psychiatric Association *Diagnostic and Statistical Manual of Mental Disorders 3rd Edition (DSM III)*. Washington DC: APA. 1980.

American Psychiatric Association *Diagnostic and Statistical Manual of Mental Disorders (4th edition) (DSMIV)*. Washington, DC: American Psychiatric Association. 1994.

Angold A, Costello EJ, Worthman CM. Puberty and depression: the roles of age, pubertal status and pubertal timing. *Psychol Med* 1998; 28: 51-61.

Baranowska B, Radzikowska M, Wasilewska-Dziubinska E *et al.* Disturbed release of gastrointestinal peptides in anorexia nervosa and in obesity. *Diabetes Obes Metab* 2000; 2: 99-103.

Basdevant A, Pouillon M, Lahlou N *et al.* Prevalence of binge eating disorder in different populations of French women. *Int J Eat Disord* 1995; 18: 309-315.

Bastiani AM, Rao R, Weltzin T. *et al.* Perfectionism in anorexia nervosa. *Int J Eat Disord* 1995; 17: 147-152.

Bjorntorp P, Rosmond R. Hypothalamic origin of the metabolic syndrome X. *Ann N Y Acad Sci* 1999; 892: 297-307.

Bjorntorp P, Rosmond R. Neuroendocrine abnormalities in visceral obesity. *Int J Obes Relat Metab Disord* 2000a; 24 Suppl 2: S80-S85.

Bjorntorp P, Rosmond R. Obesity and cortisol. *Nutrition* 2000b; 16: 924-936.

Blouin AG, Blouin J, Bushnik T. *et al.* A double-blind placebo-controlled glucose challenge in bulimia nervosa: psychological effects. *Biol Psychiatry* 1993; 33: 160-168.

Brecelj and M, Tchanturia K Sanchez P Troop N Connan F Tomori M Treasure J. Childhood temperament. 2001; (in press).

Brewerton TD, Jimerson DC. Studies of serotonin function in anorexia nervosa. *Psychiatry Res* 1996; 62: 31-42.

Brewerton TD, Lesem MD, Kennedy A. *et al.* Reduced plasma leptin concentrations in bulimia nervosa. *Psychoneuroendocrinology* 2000; 25: 649-658.

Brown N, Ward, A, Surwit, R, Tiller, J, Lightman, S, and Treasure, JL Campbell IC. Evidence for the persistence of endocrine, metabolic and behavioural abnormalities in subjects recovered from anorexia nervosa. *Int J Eat Dis*ord 2001; In Press.

Bruch H. *Eating disorders: Obesity, anorexia nervosa and the person within.* New York, Basic Books. 1973.

Bruins-Slot L, Gorwood P, Bouvard M *et al.* Lack of association between anorexia nervosa and D3 dopamine receptor gene. *Biol Psychiatry* 1998; 43: 76-78.

Bulik CM Sullivan PF Wade TD *et al.* Twin studies of eating disorders: a review. *Int J Eat Disord* 2000; 27: 1-20.

Campbell DA, Sundaramurthy D, Gordon D *et al.* Association between a marker in the UCP-2/UCP-3 gene cluster and genetic susceptibility to anorexia nervosa. *Mol Psychiatry* 1999; 4: 68-70.

Campbell DA, Sundaramurthy D, Markham AF *et al.* Lack of association between 5-HT2A gene promoter polymorphism and susceptibility to anorexia nervosa [letter]. *Lancet* 1998; 351: 499.

Channon S, Hemsley D, de Silva P *et al.* Selective processing of food words in anorexia nervosa. *Br J Clin Psychol* 1988; 27 (3): 259-260.

Cnattingius S, Hultman CM, Dahl M *et al.* Very preterm birth, birth trauma, and the risk of anorexia nervosa among girls. *Arch Gen Psychiatry* 1999; 56: 634-638.

Collier DA, Arranz MJ, Li T *et al.* Association between 5-HT2A gene promoter polymorphism and anorexia nervosa [letter]. Lancet 1997; 350: 412.

Connan, F., Campbell, I., Katzman, M., Lightman S., and Treasure, J. A neurodevelopmental model for the aetiology of anorexia nervosa. *DK* 2000; in press.

Coppen AJ, Gupta RK, Eccleston EG *et al.* Letter: Plasma-tryptophan in anorexia nervosa. *Lancet* 1976; 1: 961.

Cowen PJ, Smith KA. Serotonin, dieting, and bulimia nervosa. *Adv Exp Med Biol* 1999; 467: 101-104.

Devlin MJ, Walsh BT, Guss JL *et al.* Postprandial cholecystokinin release and gastric emptying in patients with bulimia nervosa. *Am J Clin Nutr* 1997; 65: 114-120.

Drewnowski A, Halmi KA, Pierce B *et al.* Taste and eating disorders. *Am J Clin Nutr* 1987; 46: 442-450.

Dunn AJ, Berridge CW. Physiological and behavioral responses to corticotropin-releasing factor administration: is CRF a mediator of anxiety or stress responses? *Brain Res Rev* 1990; 15: 71-100.

Eldredge KL, Agras WS. The relationship between perceived evaluation of weight and treatment outcome among individuals with binge eating disorder. *Int J Eat Disord* 1997; 22: 43-49.

Ellison Z, Foong J, Howard R *et al.* Functional anatomy of calorie fear in anorexia nervosa [letter]. *Lancet* 1998; 352: 1192.

Enoch MA, Kaye WH, Rotondo A *et al.* 5-HT2A promoter polymorphism -1438G/A, anorexia nervosa, and obsessive- compulsive disorder [letter]. *Lancet* 1998; 351: 1785-1786.

Epel E, Lapidus R, McEwen B *et al.* Stress may add bite to appetite in women: a laboratory study of stress- induced cortisol and eating behavior. *Psychoneuroendocrinology* 2001; 26: 37-49.

Epel ES, McEwen B, Seeman T *et al.* Stress and body shape: stress-induced cortisol secretion is consistently greater among women with central fat. *Psychosom Med* 2000; 62: 623-632.

Fairburn CG, Cooper Z, Doll HA *et al.* Risk factors for anorexia nervosa: three integrated case-control comparisons. *Arch Gen Psychiatry* 1999a; 56: 468-476.

Fairburn CG, Cowen PJ, Harrison PJ. Twin studies and the etiology of eating disorders. *Int J Eat Disord* 1999; 26: 349-358.

Fairburn C G, Doll HA, Welch SL *et al.* Risk factors for binge eating disorder: a community-based, case-control study. *Arch Gen Psychiatry* 1998; 55: 425-432.
Fairburn CG, Welch SL, Doll HA *et al.* Risk factors for bulimia nervosa. A community-based case-control study. *Arch Gen Psychiatry* 1997; 54: 509-517.
Ferron F, Considine RV, Peino R *et al.* Serum leptin concentrations in patients with anorexia nervosa, bulimia nervosa and non-specific eating disorders correlate with the body mass index but are independent of the respective disease. *Clin.Endocrinol.(Oxf)* 1997; 46: 289-293.
Fichter MM, Pirke KM, Pollinger J *et al.* Disturbances in the hypothalamo-pituitary-adrenal and other neuroendocrine axes in bulimia. *Biol Psychiatry* 1990; 27: 1021-1037.
Fichter MM, Quadflieg N, Brandl B. Recurrent overeating: an empirical comparison of binge eating disorder, bulimia nervosa, and obesity. *Int J Eat Disord* 1993; 14: 1-16.
Franko DL, Wolfe BE, Jimerson DC. Elevated sweet taste pleasantness ratings in bulimia nervosa. *Physiol Behav* 1994; 56: 969-973.
Frey J, Hebebrand J, Muller B *et al.* Reduced body fat in long-term followed-up female patients with anorexia nervosa. *J Psychiatr Res* 2000; 34: 83-88.
Fulton S, Woodside B, Shizgal P. Modulation of brain reward circuitry by leptin. *Science* 2000; 287: 125-128.
Garfinkel PE, Lin E, Goering P *et al.* Bulimia nervosa in a Canadian community sample: prevalence and comparison of subgroups. *Am J Psychiatry* 1995; 152: 1052-1058.
Gautier JF, Chen K, Salbe AD *et al.* Differential brain responses to satiation in obese and lean men. *Diabetes* 2000; 49: 838-846.
Gautier JF, Chen K, Uecker A *et al.* Regions of the human brain affected during a liquid-meal taste perception in the fasting state: a positron emission tomography study. *Am J Clin Nutr* 1999; 70: 806-810.
Geiselman PJ, Anderson AM, Dowdy ML *et al.* Reliability and validity of a macronutrient self-selection paradigm and a food preference questionnaire. *Physiol Behav* 1998; 63: 919-928.
Gelder MG, Lopez-Ibor JJ, Andreason, N. *New Oxford Textbook of Psychiatry*. Oxford University Press. 2000.
Gendall KA, Kaye WH, Altemus M *et al.* () Leptin, neuropeptide Y, and peptide YY in long-term recovered eating disorder patients. *Biol Psychiatry* 1999; 46: 292-299.
Gendall KA, Sullivan PF, Joyce PR *et al.* Food cravings in women with a history of anorexia nervosa. *Int J Eat Disord* 1997; 22: 403-409.
Geracioti TDJr, Liddle RA. Impaired cholecystokinin secretion in bulimia nervosa. *N Engl J Med* 1988; 319: 683-688.
Geracioti TDJr, Liddle RA, Altemus M *et al.* Regulation of appetite and cholecystokinin secretion in anorexia nervosa. *Am J Psychiatry* 1992; 149: 958-961.
Goldbloom DS, Kennedy SH, Kaplan AS *et al.* Anorexia nervosa and bulimia nervosa. *CMAJ* 1989; 140: 1149-1154.
Goldfein JA, Walsh BT, LaChaussee JL *et al.* Eating behavior in binge eating disorder. *Int J Eat Disord* 1993; 14: 427-431.
Goldner EM, Srikameswaran S, Schroeder ML *et al.* Dimensional assessment of personality pathology in patients with eating disorders. *Psychiatry Res* 1999; 85: 151-159.
Goodwin GM, Shapiro CM, Bennie J *et al.* The neuroendocrine responses and psychological effects of infusion of L- tryptophan in anorexia nervosa. *Psychol Med* 1989; 19: 857-864.
Gordon CM, Dougherty DD, Rauch SL *et al.* Neuroanatomy of human appetitive function: A positron emission tomography investigation. *Int J Eat Disord* 2000; 27: 163-171.
Greeno CG, Wing RR, Shiffman S. Binge antecedents in obese women with and without binge eating disorder. *J Consult Clin Psychol* 2000; 68: 95-102.

Han L, Nielsen DA, Rosenthal NE *et al.* No coding variant of the tryptophan hydroxylase gene detected in seasonal affective disorder, obsessive-compulsive disorder, anorexia nervosa, and alcoholism. *Biol Psychiatry* 1999; 45: 615-619.

Harty RF, Pearson PH, Solomon TE *et al.* Cholecystokinin, vasoactive intestinal peptide and peptide histidine methionine responses to feeding in anorexia nervosa. *Regul Pept* 1991; 36: 141-150.

Hebebrand J, Ballauff A, Hinney A *et al.* [Body weight regulation in anorexia nervosa with special attention to leptin secretion]. *Nervenarzt* 1999; 70: 31-40.

Hebebrand J, Remschmidt H. Anorexia nervosa viewed as an extreme weight condition: genetic implications. *Hum Genet* 1995; 95: 1-11.

Hetherington MM, Altemus M, Nelson ML *et al.* Eating behavior in bulimia nervosa: multiple meal analyses. *Am J Clin Nutr* 1994; 60: 864-873.

Hinney A, Bornscheuer A, Depenbusch M *et al.* No evidence for involvement of the leptin gene in anorexia nervosa, bulimia nervosa, underweight or early onset extreme obesity: identification of two novel mutations in the coding sequence and a novel polymorphism in the leptin gene linked upstream region. *Mol Psychiatry* 1998; 3: 539-543.

Hinney A, Herrmann H, Lohr T *et al.* No evidence for an involvement of alleles of polymorphisms in the serotonin1Dbeta and 7 receptor genes in obesity, underweight or anorexia nervosa. *Int J Obes Relat Metab Disord* 1999; 23: 760-763.

Hinney A, Lentes KU, Rosenkranz K *et al.* Beta 3-adrenergic-receptor allele distributions in children, adolescents and young adults with obesity, underweight or anorexia nervosa. *Int J Obes Relat Metab Disord* 1997; 21: 224-230.

Hinney A, Schmidt A, Nottebom K *et al.* Several mutations in the melanocortin-4 receptor gene including a nonsense and a frameshift mutation associated with dominantly inherited obesity in humans. *J Clin Endocrinol Metab* 1999; 84: 1483-1486.

Hinney A, Ziegler A, Nothen MM *et al.* 5-HT2A receptor gene polymorphisms, anorexia nervosa, and obesity [letter]. *Lancet*, 1997; 350: 1324-1325.

Hirano H, Tomura N, Okane K *et al.* Changes in cerebral blood flow in bulimia nervosa. *J Comput Assist Tomogr* 1999; 23: 280-282.

Hoek HW. The incidence and prevalence of anorexia nervosa and bulimia nervosa in primary care. *Psychol Med* 1991; 21: 455-460.

Jarman M, Walsh S. Evaluating recovery from anorexia nervosa and bulimia nervosa: integrating lessons learned from research and clinical practice. *Clin Psychol Rev* 1999; 19: 773-788.

Jimerson DC, Lesem MD, Kaye WH *et al.* Low serotonin and dopamine metabolite concentrations in cerebrospinal fluid from bulimic patients with frequent binge episodes. *Arch Gen Psychiatry* 1992; 49: 132-138.

Jimerson DC, Mantzoros C, Wolfe BE *et al.* Decreased Serum Leptin in Bulimia Nervosa. *J Clin Endocrinol Metab* 2000; 85: 4511-4514.

Jiskra J, Haluzik M, Svobodova J *et al.* [Serum leptin levels and soluble leptin receptors in female patients with anorexia nervosa]. *Cas Lek Cesk* 2000; 139: 660-663.

Johnson WG, Jarrell MP, Chupurdia KM *et al.* Repeated binge/purge cycles in bulimia nervosa: role of glucose and insulin. *Int J Eat Disord* 1994; 15, 331-341.

Karhunen LJ, Vanninen EJ, Kuikka JT *et al.* Regional cerebral blood flow during exposure to food in obese binge eating women. *Psychiatry Res* 2000; 99: 29-42.

Karwautz A, Rabe-Hesketh S, Hu X *et al.* Individual-specific risk factors for anorexia nervosa: a pilot study using a discordant sister-pair design. *Psychol Med* 2001; 31: 317-329.

Katzman DK, Lambe EK, Mikulis DJ *et al.* Cerebral gray matter and white matter volume deficits in adolescent girls with anorexia nervosa. *J Pediatr* 1996; 129: 794-803.

Kaye WH, Gwirtsman HE, George DT *et al.* CSF 5-HIAA concentrations in anorexia nervosa: reduced values in underweight subjects normalize after weight gain. *Biol Psychiatry* 1988; 23: 102-105.

Kenardy J, Mensch M, Bowen K *et al.* A comparison of eating behaviors in newly diagnosed NIDDM patients and case-matched control subjects. *Diabetes Care* 1994; 17: 1197-1199.

Kendler KS, MacLean C, Neale M *et al.* The genetic epidemiology of bulimia nervosa. *Am J Psychiatry* 1991; 148: 1627-1637.

Kinzl JF, Traweger C, Trefalt E *et al.* Binge eating disorder in females: a population-based investigation. *Int J Eat Disord* 1999; 25: 287-292.

Kissileff HR, Wentzlaff TH, Guss JL *et al.* A direct measure of satiety disturbance in patients with bulimia nervosa. *Physiol Behav* 1996; 60: 1077-1085.

Klump KL, McGue M, Iacono WG. Age differences in genetic and environmental influences on eating attitudes and behaviors in preadolescent and adolescent female twins. *J Abnorm Psychol* 2000; 109: 239-251.

Klump KL, Miller KB, Keel PK, Iacono WG, McGue M. Genetic and environmental influences on anorexia nervosa syndromes in a population based twin sample. *Psychol med* 2001; (in press).

Koch KL, Bingaman S, Tan L *et al.* Visceral perceptions and gastric myoelectrical activity in healthy women and in patients with bulimia nervosa. *Neurogastroenterol Motil* 1998; 10: 3-10.

Kortegaard LS, Hoerder K, Joergensen J *et al.* A preliminary population-based twin study of self-reported eating disorder. *Psychol Med* 2001; 31: 361-365.

LaChaussee JL, Kissileff HR, Walsh BT *et al.* The single-item meal as a measure of binge-eating behavior in patients with bulimia nervosa. *Physiol Behav* 1992; 51: 593-600.

Lear SA, Pauly RP, Birmingham CL. Body fat, caloric intake, and plasma leptin levels in women with anorexia nervosa. *Int J Eat Disord* 1999; 26: 283-288.

LeGoff DB, Leichner P, Spigelman MN. Salivary response to olfactory food stimuli in anorexics and bulimics. *Appetite* 1988; 11: 15-25.

Leonard T, Perpina C, Bond A, Treasure J. Assessment of test meal induced autonomic arousal in anorexic, bulimic and control females. *European Eating Disorders Rev* 1998; 6: 188-200.

Lilenfeld LR, Kaye WH. Genetic studies of anorexia and bulimia nervosa . In *Neurobiology in the treatment of eating disorders.* Eds Hoek HW, Treasure J, Katzman MA. 169-194. Chichester, Wiley. 1998.

Lissau I, Sorensen TI. Parental neglect during childhood and increased risk of obesity in young adulthood. *Lancet* 1994; 343: 324-327.

Lucas AR, Beard CM, O'Fallon WM *et al.* 50-year trends in the incidence of anorexia nervosa in Rochester, Minn.: a population-based study. *Am J Psychiatry* 1991; 148: 917-922.

McQuade R, Young AH. Future therapeutic targets in mood disorders: the glucocorticoid receptor. *Br J Psychiatry* 2000; 177: 390-395.

Mehler PS, Eckel RH, Donahoo WT. () Leptin levels in restricting and purging anorectics. *Int J Eat Disord* 1999; 26: 189-194.

Modell S, Lauer CJ, Schreiber W *et al.* Hormonal response pattern in the combined DEX-CRH test is stable over time in subjects at high familial risk for affective disorders. *Neuropsychopharmacology* 1998; 18: 253-262.

Monteleone P, Bortolotti F, Fabrazzo M *et al.* Plamsa leptin response to acute fasting and refeeding in untreated women with bulimia nervosa. *J Clin Endocrinol Metab* 2000; 85: 2499-2503.

Monteleone P, Di Lieto A, Tortorella A *et al.* Circulating leptin in patients with anorexia nervosa, bulimia nervosa or binge-eating disorder: relationship to body weight, eating patterns, psychopathology and endocrine changes. *Psychiatry Res* 2000b; 94: 121-129.

Mussell MP, Mitchell JE, Weller CL *et al.* Onset of binge eating, dieting, obesity, and mood disorders among subjects seeking treatment for binge eating disorder. *Int J Eat Disord* 1995; 17: 395-401.

Nakai Y, Hamagaki S, Kato S *et al.* Leptin in women with eating disorders. *Metabolism* 1999; 48: 217-220.
North C, Gowers S. Anorexia nervosa, psychopathology, and outcome. *Int J Eat Disord* 1999; 26: 386-391.
Nozoe S, Naruo T, Nakabeppu Y *et al.* Changes in regional cerebral blood flow in patients with anorexia nervosa detected through single photon emission tomography imaging. *Biol Psychiatry* 1993; 34: 578-580.
Nozoe S, Naruo T, Yonekura R *et al.* Comparison of regional cerebral blood flow in patients with eating disorders. *Brain Res Bull* 1995; 36: 251-255.
O'Dwyer AM, Lucey JV, Russell GF. Serotonin activity in anorexia nervosa after long-term weight restoration: response to D-fenfluramine challenge. *Psychol Med* 1996; 26: 353-359.
Palmer RL. *Management of eating disorders*. Chichester, Wiley. 2000.
Phillipp E, Pirke KM, Kellner MB *et al.* Disturbed cholecystokinin secretion in patients with eating disorders. *Life Sci* 1991; 48: 2443-2450.
Pinel JP, Assanand S, Lehman DR. Hunger, eating, and ill health. *Am Psychol* 2000; 55: 1105-1116.
Pirke KM, Kellner MB, Friess E *et al.* Satiety and cholecystokinin. *Int J Eat Disord* 1994; 15: 63-69.
Polito A, Fabbri A, Ferro-Luzzi A *et al.* Basal metabolic rate in anorexia nervosa: relation to body composition and leptin concentrations. *Am J Clin Nutr* 2000; 71: 1495-1502.
Polivy J, Herman CP. Dieting and binging. A causal analysis. *Am Psychol* 1985; 40: 193-201.
Rathner G, Messner K. Detection of eating disorders in a small rural town: an epidemiological study. *Psychol Med* 1993; 23: 175-184.
Ricca V, Mannucci E, Calabro A *et al.* Anorexia nervosa and celiac disease: two case reports. *Int J Eat Disord* 2000: 27: 119-122.
Robinson PH. Perceptivity and paraceptivity during measurement of gastric emptying in anorexia and bulimia nervosa. *Br J Psychiatry* 1989; 154: 400-405.
Robinson PH, Clarke M, Barrett J. Determinants of delayed gastric emptying in anorexia nervosa and bulimia nervosa. *Gut* 1988; 29: 458-464.
Rolls ET, Baylis LL. Gustatory, olfactory, and visual convergence within the primate orbitofrontal cortex. *J Neurosci* 1994; 14: 5437-5452.
Rosenkranz K, Hinney A, Ziegler A *et al.* Systematic mutation screening of the estrogen receptor beta gene in probands of different weight extremes: identification of several genetic variants. *J Clin Endocrinol Metab* 1998b 83: 4524-4527.
Rosenkranz K, Hinney A, Ziegler A *et al.* Screening for mutations in the neuropeptide Y Y5 receptor gene in cohorts belonging to different weight extremes. *Int J Obes Relat Metab Disord* 1998a; 22: 157-163.
Rosmond R, Bjorntorp P. Occupational status, cortisol secretory pattern, and visceral obesity in middle-aged men. *Obes Res* 2000; 8: 445-450.
Schmidt U, Tiller J, Blanchard M *et al.* Is there a specific trauma precipitating anorexia nervosa? *Psychol Med* 1997; 27: 523-530.
Shoebridge P, Gowers SG. Parental high concern and adolescent-onset anorexia nervosa. A case- control study to investigate direction of causality. *Br J Psychiatry* 2000; 176: 132-137.
Simon Y, Bellisle F, Monneuse MO *et al.*Taste responsiveness in anorexia nervosa. *Br J Psychiatry* 1993; 162: 244-246.
Smith KA, Fairburn CG, Cowen PJ. Symptomatic relapse in bulimia nervosa following acute tryptophan depletion. *Arch Gen Psychiatry* 1999; 56: 171-176.
Soundy TJ, Lucas AR, Suman VJ *et al.* Bulimia nervosa in Rochester, Minnesota from 1980 to 1990. *Psychol Med* 1995; 25: 1065-1071.

Spitzer RL, Devlin B, Walsh A, Hasdin D, Wing R, Stunkard AJ, Wadden T, Yanovski JA, Agras S, Mitchell JE. Binge eating disorder: To be or not to be in DSMIV. *Int J Eat Disord* 1991; 10: 627-629.

Spitzer RL, Stunkard A, Yanovski S *et al.* Binge eating disorder should be included in DSM-IV: a reply to Fairburn et al.'s "the classification of recurrent overeating: the binge eating disorder proposal". *Int J Eat Disord* 1993; 13: 161-169.

Strober M, Freeman R, Lampert C *et al.* Controlled family study of anorexia nervosa and bulimia nervosa: evidence of shared liability and transmission of partial syndromes. *Am J Psychiatry* 2000; 157: 393-401.

Sunday SR, Halmi KA. Taste perceptions and hedonics in eating disorders. *Physiol Behav* 1990; 48: 587-594.

Sunday SR, Halmi KA. M. *Appetite* 1996; 26: 21-36.

Tamai H, Kiyohara K, Mukuta T *et al.* Responses of growth hormone and cortisol to intravenous glucose loading test in patients with anorexia nervosa. *Metabolism* 1991; 40: 31-34.

Tamai H, Takemura J, Kobayashi N *et al.* Changes in plasma cholecystokinin concentrations after oral glucose tolerance test in anorexia nervosa before and after therapy. *Metabolism* 1993; 42: 581-584.

Tataranni PA, Cizza G, Snitker S *et al.* Hypothalamic-pituitary-adrenal axis and sympathetic nervous system activities in Pima Indians and Caucasians. *Metabolism* 1999; 48: 395-399.

Telch CF, Agras WS. Obesity, binge eating and psychopathology: are they related? *Int J Eat Disord* 1994; 15: 53-61.

Telch CF, Stice E. Psychiatric comorbidity in women with binge eating disorder: prevalence rates from a non-treatment-seeking sample. *J Consult Clin Psychol* 1998; 66: 768-776.

Turnbull S, Ward A, Treasure J *et al.* The demand for eating disorder care. An epidemiological study using the general practice research database. *Br J Psychiatry* 1996; 169: 705-712.

Tuschen-Caffier B, Vogele C. Psychological and physiological reactivity to stress: an experimental study on bulimic patients, restrained eaters and controls. *Psychother Psychosom* 1999; 68: 333-340.

Uhe AM, Szmukler GI, Collier GR *et al.* Potential regulators of feeding behavior in anorexia nervosa. *Am J Clin Nutr* 1992; 55: 28-32.

Wade TD, Bulik CM, Neale M *et al.* Anorexia nervosa and major depression: shared genetic and environmental risk factors. *Am J Psychiatry* 2000; 157: 469-471.

Walsh BT, Roose SP, Katz JL *et al.* Hypothalamic-pituitary-adrenal-cortical activity in anorexia nervosa and bulimia. *Psychoneuroendocrinology* 1987; 12: 131-140.

Walters EE, Kendler KS. Anorexia nervosa and anorexic-like syndromes in a population-based female twin sample. *Am J Psychiatry* 1995; 152: 64-71.

Wang GJ, Volkow ND, Logan J *et al.* Brain dopamine and obesity. *Lancet* 2001; 357: 354-357.

Ward A, Brown N, Lightman S *et al.* Neuroendocrine, appetitive and behavioural responses to d-fenfluramine in women recovered from anorexia nervosa. *Br J Psychiatry* 1998; 172: 351-358.

Watson AM, Poloyac SM, Howard G *et al.* Effect of leptin on cytochrome P-450, conjugation, and antioxidant enzymes in the ob/ob mouse. *Drug Metab Dispos* 1999; 27: 695-700.

Westen D, Harnden-Fischer J. Personality profiles in eating disorders: rethinking the distinction between axis I &axis II. 2001; (in press).

Whitehouse AM, Cooper PJ, Vize CV *et al.* Prevalence of eating disorders in three Cambridge general practices: hidden and conspicuous morbidity. *Br J Gen Pract* 1992; 42: 57-60.

Wilfley DE, Friedman MA, Dounchis JZ *et al.* Comorbid psychopathology in binge eating disorder: relation to eating disorder severity at baseline and following treatment. *J Consult Clin.Psychol* 2000; 68: 641-649.

Wilfley DE, Schwartz MB, Spurrell EB *et al.* Assessing the specific psychopathology of binge eating disorder patients: interview or self-report? *Behav Res Ther* 1997; 35: 1151-1159.
Wilson GT, Fairburn CG. Cognitive treatments for eating disorders. *J Consult Clin.Psychol* 1993; 61: 261-269.
Wise RA, Rompre PP. Brain dopamine and reward. *Annu Rev Psychol* 1989; 40: 191-225.
Yanovski SZ, Nelson JE, Dubbert BK *et al.* Association of binge eating disorder and psychiatric comorbidity in obese subjects. *Am J Psychiatry* 1993; 150: 1472-1479.
Zhao M, Liu Z, Su J. The time-effect relationship of central action in acupuncture treatment for weight reduction. *J Tradit Chin Med* 2000; 20: 26-29.
Ziegler A, Gorg T. 5-HT2A gene promoter polymorphism and anorexia nervosa [letter]. *Lancet* 1999; 353: 929.
Zobel AW, Nickel T, Sonntag A, Uhr M, Holsboer F, Ising M. Cortisol response to the combined dexamethasone/CRH test as predictor of relapse in patients with remitted depression: a prospective study. *Journal of Psychiatric Research* 2001; (in press).

Chapter 3

FREE-CHOICE DIET SELECTION -- THE ECONOMICS OF EATING

Carl V. Phillips
Minnesota Center for the Philosophy of Science
University of Minnesota
746 Heller Hall
271 19th Ave. S.
Minneapolis, MN 55455

Abstract Dietary choice is a fundamentally economic phenomenon in a social context. Only by understanding the socio-economic perspective can we understand rational free-choice dietary choice, which in turn is necessary for truly understanding departures from rational choice. This chapter presents an overview of economic choice and how it applies to dietary choices, good and bad, and draws upon several non-traditional viewpoints of intertemporal choice. Methods exist for modeling some of the complexity of dietary choice. Laboratory modeling of dietary choice will not be effective (or ethical) without consideration of which economic and social phenomenon the model captures.

INTRODUCTION

In the clinical world, dietary choice is often addressed as an extension of the physiology of nutrition or the health impacts of nutritional intake. When the focus of inquiry is the food itself or the body of whoever eats it, then it is natural to make the act of choosing secondary to the physiology. When clinical science expands its attention from the physiological to the psychological, the emphasis is often the dark side of free choice: concern about eating disorders. The non-pathological aspects of preferences, tradeoffs, and choices are often minimized. In many cases, this makes sense. When someone is engaged in acute self-destructive behavior, an analysis of how people normally choose their diets is

J.B. Owen et al. (eds.), Animal Models – Disorders of Eating Behaviour and Body Composition, 51–66.

probably not very useful. In most cases, though, individual choice cannot be properly understood using a medical or biological paradigm as a starting point.

Dietary choice in free-living modern humans is not primarily a physiological phenomenon or a matter of psycho-pathology. Eating is a fundamentally social and economic phenomenon, and it is important to frame our thinking about food and nutrition in that context. This is especially true when we are attempting to model people's decisions about diet. Whether the model is a simple conceptual sketch, a complicated computer simulation, a study of human subjects in a controlled setting, or an animal model, it will be of limited value if it attempts to describe behavior without the social and economic context. Physiology can often be modeled effectively with people treated as machines, but behavior is difficult to model accurately without a recognition of humans as complex social actors.

This chapter takes a view of dietary choice based on social science (a term which is this context is intended to be exclusive of behavioral science), primarily the view from economics. The value -- indeed, critical importance -- of economics in analyzing and understanding public health phenomena, like eating behavior, is often overlooked. In part, this seems to be due to a popular misconception (including among some economists) that economics is merely the study of finance and commerce. Economics is better defined as the science of tradeoffs among scarce resources, and the subfield of welfare economics focuses on those decisions in the context of individual human welfare. Choices about which foods to eat, the evolution of a society's eating patterns, and the tradeoffs between good health and unhealthy urges can be informed by welfare economics and other social sciences.

This chapter does not attempt to provide an overview of the specific choices people do make about food, either rationally or irrationally, in a particular society at a particular time. That information is covered in a vast literature that is beyond the present scope. Usually missing from such analyses is an underlying framework of individual choice beyond observing associations in the data. The present chapter attempts to lay out that framework, presenting the basic economic view and then drawing on selected economics-based theories for further elaboration.

This chapter proceeds as follows. The basic economic model of free-choice decisions is presented and the difficulties in modeling such decisions in all their complexity are discussed. This framework is then applied to dietary decisions, first as individual choice and then as part of a social context. A brief overview of a computer model for capturing some of this complexity is presented. Finally, irrational dietary choices are looked at from a few economic angles, including the competition between intertemporal goals and addictions and related concepts.

It should be noted that the author does not claim any significant expertise in clinical-level eating disorders or lab research. The emphasis here is on health and unhealthy eating patterns in a social context, but not on behavior that represents a total breakdown of individual rationality. Clinicians and researchers who focus on pathological conditions must determine what lessons can be carried from this to their areas of expertise.

MODELING ECONOMIC DECISION MAKING

The Economics of Free Choice

The basic welfare economics model of decision making can be summarized by saying that people try to *maximize discounted net expected benefits* subject to their *constraints*. This simple concept (which is expanded upon below) is usually complicated to operationalize and sometimes even out-and-out wrong empirically, but it is an important starting point and needs to be understood. If people appear to be making bad choices, it is not enough to declare the choices bad and try to intervene. It is important to understand exactly what "bad" means, how someone's choice is deviating from optimal behavior (if at all), and why she is making that decision. Economic models, with their assumptions about rationality, good information, and people generally having an idea of what is best for themselves, are often wrong, but they should be given rebuttable presumption. That is, until we are given a good reason to assume people's decisions are not in their best interest, we should treat them as if they are, and when we become convinced that they are not, we should try to explain why. In modern Western societies (and, indeed, in a world where liberal individualism and democracy are being declared victors in the war of political philosophies), we are on shaky ground if we start with other assumptions in research or clinical practice.

The above phrase "discounted net expected benefits" contains a large number of assumptions about individual choice, some undeniably reasonable and some frequently false (though convenient). "Benefits" is generally taken by economists to include anything that someone wants, a philosophy that is sometimes labeled "preference utilitarianism." In particular, economics takes pains to not judge one source of benefit as more real than another (with a few exceptions, mainly preferences that severely violate someone else's rights like a penchant for committing battery). This means that a desire to eat particular foods or quantities, the pursuit of bodily appearance, and other consumption decisions are not subordinate to a desire to be healthy (or an indifference about it). Without a specific argument for why a particular preference ought not to be accepted (e.g., it is caused by a brain chemical imbalance that renders someone incapable of making rational decisions), the liberal ethic does not deny someone the right to have it.

The economic notion of what is good (and thus the resulting policy analytic goals) takes as given the individual preferences and intensity thereof, and defines the social good based on them. From a normative perspective, this narrow definition of value is controversial and often problematic -- important considerations like obligations and justice are absent. Furthermore, it ignores the "path dependent" nature of preferences in a social setting, a topic addressed below. As a description of the motives that lead to individual behavior, though, it is quite effective, both normatively and positively. Even when it is not an accurate description of behavior, asking *why* it is not accurate is an important step in trying to deal with the behavior.

The modifier "net" in "net expected value" refers to the need to add up all benefits and subtract all costs. (There is no fundamental difference between costs and benefits to someone, other than defining a starting point. A cost is just a benefit lost or foregone.) The modifier "expected" says that decisions should be made based on the expected value of outcomes, that is, the probability-weighted average (mean) of possible net benefit realizations. In cases where a decision's outcome is largely dependent on uncertainty about facts or realizations of the future state of the world, this becomes crucial, but for present purposes it is not central.

"Discounted" refers to the tradeoff between present benefits and those in the future. This is a crucial aspect of health-affecting choices, and the one most often invoked (usually implicitly) to justify trying to change individual decisions. The claim that someone is not paying enough attention to his future health in making his current decisions is a statement that his discount rate is too high (i.e., when he adds present and future welfare, he divides the latter by too large a conversion factor). As both a theoretical optimum and an empirical observation, people consider a benefit or cost in the future to be worth less than the same benefit or cost now. There are certain rules that intertemporal choices should follow, but often do not, a topic that is taken up below.

The economic theory of behavior says that people will calculate net expected benefits as best they can (presumably pretty well) and make the decision that has the highest payoff. The invisible hand theorems (first alluded to by Adam Smith and more formally proven subsequently) go further, showing that if people and the economy have certain reasonably realistic characteristics, this will lead to an optimal outcome for everyone. In reality, these characteristics are reasonably accurate, but not completely.

To the extent that individuals make rational optimizing choices about their own consumption, the liberal economic argument goes, policy makers and clinicians have no business trying to change their behavior (except to protect others from negative spillovers from it). After all, we in public policy and public health justify our interventions based on improving people's lives. So barring some compelling reasons to the contrary, interventions should only be made when they improve someone's welfare. An immediate corollary of this philosophy is that when we declare someone's free choices to be poor and intervene to change them, it is incumbent upon us to understand what a good free choice would look like and what violation of optimizing behavior is leading to the poor choice. Otherwise it is hard to honestly assert that the behavior should be changed. In the case of dietary choice, we might envision poor choices -- and thus a justification for intervention -- resulting from social phenomena or individual departures from rationality.

The Challenge of Modeling Economic Behavior

Starting with the socio-economic model of individual welfare optimization, it is immediately apparent that modeling preference and choice is difficult. A model needs to be a simplification that preserves relevant aspects of the system being studied. The more complicated the system, the more difficult it is to create a sufficiently complete model. This is one reason why so much research in public health and other fields focuses on observational studies of natural behavior in the real-world. Other research uses experiments, but restricts them to interventions within the complicated system that is the real world, trying to leave all else in the system unchanged.

It is widely accepted in health research that human physiology is sufficiently complicated that a cell-culture or other species will provide some information about some questions of interest, but a relatively poor picture of the overall system. Some bio-medical models will be better than others, and a lot of work goes into determining the potential and limits of such simplifications. What is often overlooked is that socio-economic systems are as complicated as human bodies.

Modeling often leads to the temptation to substitute precision for accuracy. A precise measurement (e.g., from being able to repeat a cheap controlled laboratory study enough times to get a very narrow confidence interval for whatever is being measured) of something we do not actually want to know is inferior to a rough cut at what we are actually trying to learn about. It is seldom an effective epistemic strategy to engage in "lamppost" modeling or research (a reference to the joke about the drunk looking for his keys under the lamppost, despite having lost them elsewhere, because the light is better). Using a convenient simple model of choice creates a risk of doing lamppost research. The rarified atmosphere faced by an animal in a cage (or humans in a psychology lab) creates a risk of losing the key economic and social elements of choice. This limitation applies to a wide variety of socio-economic research done with such models, including dietary choice.

Choice in a natural setting involves many complications that are difficult to simulate in a laboratory setting. These include:

- the existence of more options than can be effectively considered by participants, requiring the use of simplifying heuristics,
- individual and cultural habits that distract from rational choice,
- competing challenges and goals that limit the attention devoted to any given decision,
- complicated goals that combine hedonic pleasure, investment considerations (including long-term health), and social obligations,
- limited knowledge of options and the full impact they will have.

Modelers need to explicitly consider which of these aspects of real-world free choice are lost in their models and whether that makes a difference. Animal models face this challenge as well as serious ethical concerns.

Researchers using animal models face at least two layers of ethical concern. The first is justifying the use of non-consenting sentient beings, human or otherwise, in research at all. It is not clear that any coherent justification for this has ever been offered -- let alone been made compelling -- short of dismissing all liberal or Kantian notions of individual rights or obligations, and accepting a purely consequentialist ethic wherein the ends can justify the means. The second layer of ethical concern, even after accepting the dubious ends-justify-means ethic, is showing how the ends of a specific research project justify the means. The burden imposed on the study subjects must be justified by the knowledge that comes from the research. If the model is not accurately representing what we wish to learn about, then it seems very unlikely that such a utilitarian calculation will favor the research. If the model is basically sound, it still has to produce sufficient useful information (or, more precisely, since the decision about the research is made before this information is known, the ex ante expectation of the value of the information produced must be sufficient.

(A different type of ethical concern lies in how the research results are represented. If the model is not an effective representation of the systems about which conclusions are being offered, there is an ethical failing when misleading claims about its implications are made. Given the current norm of reporting a large portion of health research directly to the public in an unfiltered state, or to clinicians or other agents who also cannot understand all the details, this ethical consideration is particularly significant.)

With these cautions and caveats about modeling established, we can sketch a framework for analyzing human dietary choice.

INDIVIDUAL DIET CHOICE IN A SOCIAL CONTEXT

Individual Decisions, Idealized and Realistic

The basic framework for individual decisions has already been alluded to. People are faced with a menu of options (given their constraints) and make the choice that best trades off competing desires. Someone can buy a better car, eat out more, or take a nice vacation, but only one of them. She can take a higher paying job and enjoy more consumer goods, or take an easier job and enjoy nonpecuniary benefits. For simple choices of foods, the competing goals primarily include nutrition, taste and other aesthetics, availability and convenience, and monetary price (or whatever currency might be traded in the artificial economy of a laboratory, like time or effort).

When faced with the real-life choice among many foods, the economic theory says that an individual considers the multiple dimensions of each option, implicitly converting them into a scalar function that measures the net utility contribution of the choice, and chooses the option with the highest utility score. When we observe someone's actual choice, we are seeing *revealed preference*, the deeds that speak louder than any words can about which of the available choices is best. The exact preferences and tradeoffs underlying that choice can only be partially known from the single observed choice. We try in various ways to learn more, such as observing the choice between precisely defined binary options in lab settings, or making advertised changes the prices of a few goods in a retail setting to observe changing consumption patterns.

For example, when faced with a choice between eating at a local fast food restaurant and cooking and packing a lunch that more closely adheres to dietary guidelines, a person will weigh the convenience of the fast food against the nutritional advantage of the homemade meal, adding in her taste (for one or the other), the relatively prices, the time and effort spent on each, and possibly other considerations. When the calculation is done, one of the two options will emerge as superior in the eyes of that consumer at that time, and she will choose it. Multi-dimensional comparisons like this are exactly the decision making process that we hope that people use when buying most consumer products or even making major lifecycle decisions. The fact that those of us who are concerned about the public health get a little uneasy about people's rational choice of unhealthy options does not change this idea. Frequently, the fast food really is the best choice of the available options, given that people define their own "best."

The perfect optimization model of choice, despite being implicitly used in many analyses of consumer behavior, is not terribly realistic. Why do we lay out the model only to knock it down with this concession? Because it is not a bad starting point for economic analysis, because it gives us a place to base our models, and most importantly because it illustrates where model choices might go awry. The public health model of health-affecting consumer choices frequently gives consumers too little credit (when denying that their free choices are good choices) and too much credit (when implying that they make health-affecting decisions in the real world with as much consideration as lab subjects or expert panels do). Even ignoring the social context, people do not make decisions by carefully comparing all available options. The theoretical optimum is constrained by not knowing the full list of options, limited understanding of quality measures (e.g., really understanding what provides good nutritional value), and a rational unwillingness to invest the time to do such an extensive comparison, even it were possible. Instead of seeking a global optimum, consumers instead use rules of thumb, ignore many dimensions of quality, engage in *satisficing* (considering and discarding available options, in no particular order, until an options is found to offer at some minimum level of utility, at which time the search is terminated) or other simplifications.

Returning to the case of what to have for lunch, someone may have a few dozen fast food options, and she certainly has thousands of choices about what to bring from home. But we restrict ourselves to a few possibilities for each, occasionally adding a new one or forgetting an old one, since we have many other things we would like to spend our time thinking about. When we consider those options, we will use rules of thumb (vegetables are good for nutrition; large serving of meat are bad) rather than remembering the details of how the payoffs run. We might even fail to pay attention to how long it takes to cook compared to how long it takes to walk to the restaurant. Despite these limitations, our methods of *bounded rationality* based decision making usually result in pretty good choices considering the many decisions that people have to make every day. But they are much more difficult to model than are optimizing choices. Indeed, most subfields of economics have been struggling for two decades to figure out how best to treat bounded rationality in their models.

Lifestyle and Social Context

Choices of dietary patterns go beyond preferences about the inherent characteristics of particular foods. They include preferences about how much to consume, when and where, and social factors that create preferences (through history and peer effects) and directly influence choice apart from preferences (by controlling what is available or acceptable). To a large extent, these are governed by social processes.

Individual choices within this context can still be modeled as individual optimizations. Indeed, some social scientists (including this author) argue that the ideal model of social phenomena builds up from a collection of individual choices, an approach that may show phenomena that do not exist at the aggregate level and that avoids positing intangible artificial social constructs. Others argue that modeling only individuals dismisses constructs that really do exist, and that the community is more than a collection of individual decisions. Since most all of us would agree that both the individual preference and the collective community behavior matter, choosing among the social models need not be based on complicated philosophy, but simply which aspects of the behavioral process are of most interest.

At the other end of the spectrum from the individual choices already discussed is the view that the individual as a mere piece of a society, and his decisions primarily reflect the culture. The culture is the proper unit of analysis, in the extreme version of this viewpoint, not the individual. An application of this notion (or perhaps a parody or an abuse of it) can be found in the notion of excusing the individual by "blaming society" -- for everything from criminal behavior, to a upward trend in overweight, to clinical eating disorders. A more charitable example is challenging the notion that consumers can make rational choices when immersed in a materialistic, advertising-saturated, appearance-centric culture. At a less dramatic level, the decision to have fast-food for lunch only makes sense in the context of a society and commercial environment that considers fast-food to be normal and makes it easily available. Dietary pathologies become a flaw in what we as a society offer and demand of people. For this level of analysis, there is little point in trying to model eating behaviors by looking at an individual (be that an individual person, animal, or optimization formula).

The purely social viewpoint overlooks the question of how the society came into being and got where it was (a collection of individual decisions and actions) and whether context overrides preference and choice (given that every choice is made in some context, and yet the choices are still interesting). But ignoring the complex interplay between society and individual preference would be akin to ignoring the organism when studying molecular biology. It is often fine if you only want to know about the local chemistry, but it is frequently not very useful is understanding how the cells function within the whole system. No matter how promising the lab results, pharmaceuticals are ultimately measured in the context of the whole system. This is not to deny the value of chemical tests of pharmaceutical agents or the modeling of individual choices to understand dietary choice. Both are useful and perhaps necessary. But care must be taken to recognize the limits of that step and to make sure the study is limited to what it can actually measure.

Complex Systems Computer Modeling

It is possible to create a model that combines elements of individual dietary choice and the resulting commercial and cultural context. Previously (Phillips, 1999), I developed a spatial model of dietary choice which simulated the social context through complex systems programming. While this model does not offer a direct application to eating disorders, it is a first step toward modeling the complexities of free-choice dietary behavior.

Complex systems is a family of simulation techniques that look at the behavior of multi-part systems by calculating the behavior of individual parts. For example, a complex systems model of aerodynamics might look at the forces acting on each molecule in the system as time progressed. Each molecule is influenced by a variety of factors, including the position and velocity of the other molecules. As time passes, the molecules move and continue to influence each other. Ultimately, the behavior at a macro-level, that perhaps is not fully understood at the aggregate, is built from the bottom up.

My impetus for developing a complex systems model was the observation that the literature about dietary choice identified about a half dozen key motives consumers tradeoff in their choices, but most analyses described a tension between two of these. Some analyses only recognized one motive, but very few analyzed more than two (though they occasionally acknowledged them), despite their presence elsewhere in the literature. A nonsystematic review of the literature in economics, sociology, clinical dietetics, business, and evolutionary biology revealed dietary choice motives that could be categorized as:

- basic aesthetic desires (taste) and ease of preparation and consumption,

- healthfulness,
- availability, convenience of acquisition, and price,
- individual and family habits (past choices),
- a vaguely defined concept of "culture".

The first four of these were modeled as inputs into the complex systems model, with a tension between taste and healthfulness, behavior of suppliers determining availability and price (and suppliers, in turn, following the preferences of consumers), and the interaction between people and one's own history determining influence of habits. In the model, each consumer makes her own choices about consumption every period, influenced by these factors, particularly including her own past choices and the patterns of choice around her. The suppliers similarly react to the changing markets (which they also influence). The result is the emergence of self-reenforcing patterns of choice. Thus, the fifth motive found in the literature, culture, is much better described as a result (an emergent property) of the other motives, rather than a motive in itself. Indeed, different groupings that could be called cultures "evolve" from the other motives in the complex systems model.

The main value of this analysis in the present context is as a reminder of the many dimensions that constitute free-choice dietary selection in a social context. It is difficult to simplify much below this multifaceted model and still accurately represent dietary patterns. Fortunately, as the model illustrates, it is possible to build social context into a model if it is relevant.

The results of the complex systems model also illustrate the existence of many stable states. Small difference in initial conditions or small random elements in the model can result in the evolution of very different dietary "cultures" in the long run. This phenomenon, known as *path dependence*, is a reminder that any society's current foodways are largely an accident, and we could just as easily be eating very differently (perhaps more healthfully and with less tendency to eating disorders).

One tantalizing suggestion from the model is a way to change non-pathological but non-optimal dietary patters. When separate stable cultures form, each one can be thought of as a mountain top, populated by people who are trying to climb higher. Greater height represents higher utility from one's dietary pattern. Some mountains are higher than others, but if your strategy for getting higher is just to climb, you will never depart from your current choice; getting to a higher peak requires passing through a valley, and you cannot do that by myopically climbing. The valley represents dietary patterns that are difficult to maintain because they do not fit any well-established and well-supported pattern practiced by those around you. Any dietary change strategy that involves an incremental change (e.g., "I will give up 5% of the foods I am trying to avoid each week until I reach my goal") leaves you in the valley, with your urge to climb (i.e., improve your utility by making incremental changes) pulling you back toward your old mountain top.

For example, if you are trying to give up your beloved five-meals-per-week fast food habit, it might be quite difficult to achieve the first step of cutting that to four per week. ("Yesterday I decided I was going to skip fast food today, but I really don't feel like it, so I will do the day after tomorrow instead.") The solution might be to leap to the alternative clear mountain top of never eating fast food, where there is no room for fooling yourself about backsliding. If the new mountain really is higher, you may soon find that the transition was unexpectedly painless and you have no desire to go back to where you where. The true effectiveness of such a strategy is an empirical point, of course, but what evidence there is suggests that it is plausible. This metaphor might also be useful for pathological choices, since any inferior consumption pattern can be though of as being stuck at a place where your strategy for behavioral adjustment (metaphorically, climbing the hill) do not allow you to improve your choices, and a large wrenching change is frequently the clinical recommendation.

IRRATIONAL DIETARY CHOICES

Having observed the theoretical optimal food behavior and how accidents of history can result in different outcomes, we can now look at apparent fundamental departures from rationality. Some of these defy any rational explanation, though others, like addictive behavior or not worry about the future can be described in economic terms (though ultimate are unlikely to be defensible as rational).

Fundamental Breakdowns of Rational Choice

All but the most hard-core libertarian economic theorists would quickly concede that some human behavior is simply wrong. None of the rest of this analysis should be interpreted to apply to people who are suffering deep psychosis. For reasons beyond the realm of social science, sometimes choices do not match (self-perceived) preferences. Someone intentionally acts contrary to what he regards as his best interest. Or he has a disastrous collection of preferences that we are unwilling to let him indulge. Setting aside the dilemma of what is right to do for/to/with someone facing that difficulty, we can agree that something other than economics is at work. Someone who is refusing sufficient nutrition and starving herself to death is not making a free choice of her best available diet by any reasonable interpretation.

Some such pathologies may be particular amenable to modeling. A non-human animal might exhibit similar breakdowns of rational choice, and these might have fundamental commonalities with humans'. If we are interested in brain chemicals or existing learned habits rather than genuine preferences, the social context becomes less important. However, it is important to distinguish a complete breakdown of the process that translates genuine enlightened self-interest into preferences and then into choices, on one hand, and a mere error in economic calculation or even a good calculation that is just not recognized by outside observers. A few of these latter situations are taken up in the following sections.

Competing Intertemporal Goals

If most people make basically rational choices about food, why is the number of unhealthily overweight people climbing so fast? Since people have a choice about changing their behavior to avoid overweight, there are two possibilities for any individual: he is making a poor choice about current consumption and energy expenditure (exercise) and irrationally hurting his health, or he is making a rational choice to reduce his health in favor of other concerns. In other words, he is sacrificing long-term interests (becoming increasingly overweight) for short term interests (not restricting energy intake or increasing expenditure), perhaps sensibly and perhaps not. This is not fundamentally different from the choice to spend money now rather than saving it for retirement. Sometimes spending now is a good choice and sometimes it is a shortsighted mistake, depending on what we are giving up, what we are getting, and what those are worth to us.

As an aside, it is worth pausing to remember why it is that people have the desire to eat more than is healthy (and possibly as a result why other dietary pathologies have emerged). During evolutionary time, the urge to eat as much as was available when it was available was a positive adaptive strategy. Our ancestors were the individuals with stronger urges to eat when they could, and thus they survived more frequently than their contemporaries who have no descendants alive today. Evolution does not care about displeasure, only survival, so selection favored an unfulfillable hunger for more food. Our bodies and desires were not "designed" to exist in a world where that hunger can be satisfied completely and any time (there was no evolutionary pressure to perform in a certain way under those circumstances because those circumstances did not exist when we were evolving). Thus, whatever happens when we are faced with unlimited food and an opportunity for long life is an evolutionary accident (sometimes called a "quirk"). We might have been lucky, and the novel environment might have been harmless. It turns out, of course, that it is not. Many of us have a strong short-term preference to consume more energy than is healthy when presented with the opportunity. We have made a world that is much better for our health than the "natural world," but a few aspects of it present very difficult health challenges.

Eating more than is optimal for long-term health is often, perhaps usually, rational. There is some evidence that a lifetime restriction of energy well below what is considered normal and healthy will increase longevity. Very few people have decided that such a trade is worthwhile. When someone goes even further and decides to increase his risk of diabetes, heart disease, and other morbidities by eating more, he might be someone who gets so much pleasure from eating or so much pain from resisting the urge to eat that he is making the right choice.

It is common to project preferences onto him ("you can still eat a satisfying meal without it being so caloric" or "exercise is not a burden; it is very enjoyable"), but not necessarily justified. Put another way your preferences (let along a lab animal's) might not be a good model for his preferences. An economic approach reminds us that people have different tastes.

The most common justification for pushing someone to change eating behavior is an implicit claim that his discount rate is too high. That is, he is so caught up in the moment that he is failing to give proper consideration to what he is costing himself in the future. This is frequently a reasonable claim. While everyone is entitled to choose their discount rate, empirical economics has pretty good information on what people's discount rates tend to be when they fully understand and consider their choices. Overeating behaviors are difficult to explain as a discounting phenomenon based on such numbers.

An alternative description of intertemporal tradeoffs in the social science literature seems more useful for behaviors like eating, drug use, and work vs. leisure tradeoffs. Schelling's (1984, pp. 83-127) notion of "multiple selves" is one of the most underutilized models for understanding behavior in the context of the choices that really matter most to people. (It should be noted that its formal development is very limited, and it is generally treated as a thought experiment rather than a formal model, so the fault lies with economics, which has neglected it, more than those outside the field who do not pick up on it.)

Schelling models the intertemporal tradeoff not as a rational decision process between what you want now and what you will want later, nor as the foolish failure to properly weight the future. Instead there is a struggle between different individuals -- the you that exists now and the you that will exist at some point in the future -- whose welfare is related but not the same. In the classic example, before going to bed, someone puts the alarm clock on across the room on a high shelf out of concern for her future self that will arrive at work late if the alarm is not obeyed. It is the self in between, the one that will want to shut off the alarm and go back to sleep, whose preferences she is trying to thwart. Standard economics does not allow for any substantial difference between the interests of your current self and the one a mere eight hours later. Human experience clearly disagrees.

Returning to the choice of lunchtime fare, the self that exists a couple of days before the fact might plan on eating a healthy, home-cooked meal on a later day. The night before, though, preparing the packed lunch for that distant future occasion seems like too much trouble, so the job is handed off to the morning-self. That self, however, decides that an extra tap on the snooze button is worth more than packing lunch. So, finally, the lunchtime self has to spend twenty minutes going out to get a fast food meal he knows is inferior to the one he would have carried. A more dramatic example can be found when the sated compulsive eater vows to not indulge in a snack for the rest of the day, but the self a few hours later does not feel bound by that vow in the face of a compelling desire to eat. The third self that exists a few minutes later pays the price in the form of frustration and self-loathing.

This pattern is nothing new to anyone who studies dietary behavior. But the conceptualization in terms of multiple individuals making basically sensible selfish decisions, rather than a unitary actor who is fundamentally irrational, is useful. It offers ways that the tools of economics (and perhaps even management sciences) can be used in analysis and intervention. In other words, even apparently ill-chosen dietary behavior is not unlike other economic behavior. It might even be described as standard human economic behavior, only more so.

How can we model intertemporal tradeoffs about diet in a lab or experimental setting? For simple discounting decisions (two pieces of candy now or three in an hour) it is easy, though it is difficult to understand how to interpret the results. For long-term tradeoffs, we are stuck with either subjects who are very expensive to experiment on for long enough (humans) or ones who do not think about long-term goals in the same way we do (other animals). It is not clear how to begin to model the Schelling multiple selves concept, probably one of the most important tools for understanding a wide range of behaviors (not to mention for ethically justifying many medical interventions -- such as restricting someone's freedom to save her future selves from her present self's acute eating disorder -- and other paternalistic acts). Laboratory animals are sentient, feel pleasure and pain, and can understand their current options and frustrations, but they seldom possess the foresight and ability to reason about counterfactuals that is necessary for truly understanding the future self dilemma facing free-living humans. To repeat the key message about modeling, a given model may be described in terms of some phenomenon of interest, but it is necessary to think through whether it is really capturing the driving factors.

The Economics of Addition, Compulsion and Habit Formation

Instead of trying to create a bridge between the rational and the irrational, as the multiple selves concept does, some social scientists try to recapture the apparently irrational as wholly rational. Addictive behaviors are frequently cited as evidence that the rational-human based economic models cannot be accurate. In response, Becker and a series of collaborators (see, e.g., Becker, Grossman, and Murphy 1991) developed a model of rational addiction. In short, addiction is a consumption pattern that is affected in an unusual way by past consumption. Normally, past consumption reduces your desire to consume more (you get bored of going to the movies every weekend and want to do something else). In the case of addiction, past consumption, labeled *addictive stock* by Becker et al., increases your future desire to consume by increasing the value of each unit of consumption. Unfortunately, for negative addictions (i.e., recreational drugs and most other things we think of when we consider addictions), your present and future welfare is lowered by having a higher addictive stock. In other words, each bit of new consumption offers more gross benefit than it did before the past consumption took place (the relief from the first cigarette of the day is greater than the rush from the tenth cigarette of your life), but the past consumption lowers overall welfare. Thus, that high-value present consumption is just helping you climb out of a hole. This is a formalization of "running fast to stand still," and seems to be the only precise and complete definition of addiction that exists. (Biochemical definitions of addiction limit the concept to a small subset of behaviors that seem to be addictions, while the popular use of the term is very inconsistent.)

Engaging in an addictive behavior can be rational, then, when an individual anticipates the effect of present consumption on future desires and welfare and chooses a level of present consumption accordingly, ultimately making the decision about whether to ever start the addictive behavior based on whether it will have a net positive or negative impact. Few people who know and understand this theory genuinely believe that people behave nearly as rationally as Becker et al. posit. After all, to identify only the most obvious example, most smokers become addicted before an age where we consider people capable of making rational decisions about the rest of their lives. But the formal definition of addiction and the possibility that it can be rational, at least in the ongoing behavior if not the choice to start, are still very useful. The latter is useful because it denies the attitude, common in public health, that if something is considered an addictive behavior then it defies all rationality and an intervention to alter people's behavior is clearly justified.

The formal definition is useful because it helps distinguish addiction from mere habit or compulsion. Habit could be considered a situation when past actions increase value of doing the same thing again (or, identically, increase the cost of doing something else), but there is no loss of welfare from an addictive stock. Having the same meal for breakfast every day is likely a habit. Eating consistently too much might be an addiction. Both of these contrast with a compulsion, which could be described as a very strong immediate desire to engage in a certain behavior, not driven by how much you have done it in the last day or week. Indeed, if this week's consumption reduces your desire to consume more, as it might for a particular favorite food that you are temporary tired of, we have the normal economic behavior of diminishing returns, the opposite of addiction.

Why should we care about economic definitions of common terms, terms that often have different technical definitions in a medical context? Because it helps us understand the different nature of several strong desires that are often lumped together. A particular pattern of dietary choice that is clearly contrary to an individual's best interest might be best described as compulsion, addiction, or possibly merely habit. Any of these can be modeled as rational choice, multiple selves, or complete departures from rationality.

Laboratory modeling will be much more effective if a conceptual model of exactly what is happening is developed first. In many cases, when the relevant aspects of the free-choice behavior are not understood, the laboratory modeling holds little promise for being useful. The physical environment associated with a particular behavior can reenforce habit, which may be useful or it may be confounding. Similarly, an addictive behavior is not well modeled by a learned compulsion that is not actually an addiction. These observations can only be given in generalities in the present context -- it is up to individual researchers to apply them to particular work.

CONCLUSIONS

The goal of this chapter was to provide an overview of some basic concepts of dietary choice. While many aspects of the analysis may not apply in truly pathological cases, there may be more rational free-choice elements in the behavior than is generally acknowledged in a lab or clinical setting. Furthermore, it is useful to know what the pathology is departing from.

Given a general social preference for letting people make their own choices, it is important to be able to justify intervening in people's choices. It is difficult to offer such a justification without understanding what a good choice would look like and offering some explanation for why the actual choice differs. Many apparently poor choices can be explained in terms of rational behaviors. Others can be categorized as one of several well-known departures from perfect rationality. In the case of dietary choice, all choices, good or bad, need to be understood in a social context and as one element of the thousands of things people must decide about in their lives.

Finally, all researchers must consider their ethical obligations to society in terms of not wasting valuable resources and not making misleading claims. Health researchers have special obligations because their results are often rapidly translated into behaviors or policies that have major effects on people's lives. Researchers who perform experiments on humans or other animals have further ethical obligations because of the negative results their actions have on their subjects. Among the necessary conditions for fulfilling all these ethical mandates is making sure that the model being used captures and accurately represents the relevant aspects of the problem being modeled.

REFERENCES

Becker GS, Grossman M, Murphy KM. Rational Addiction and the Effect of Price on Consumption. *Am Econ Rev* 1991; 81(2): 237-41.

Phillips CV. Complex systems model of dietary choice with implications for improving diets and promoting vegetarianism. *Am J Clin Nutr* 1999; 70: 608S-614S.

Schelling, TC. *Choice and Consequence*. Cambridge: Harvard University Press. 1984.

PART 2

Diet selection and aberrations of body composition

CHAPTER 4

DIET SELECTION IN WILD ANIMALS

Lennart Hansson
Department of Conservation Biology
Swedish University of Agricultural Sciences
S-750 07 Uppsala, Sweden

ABSTRACT Animal feeding can be positioned in a gradient from roughage such as bark and old grass (detrivores-folivores) to meat-eaters (carnivores), over, e.g., fungivores and granivores. Humans occupy a position between granivores and carnivores. Granivores and carnivores are not able to eat food with thick plant cell walls and are thus restricted in their options during severe seasons. The wood mouse is considered as a wild equivalent of humans and its seasonal and annual variation in food is described. Animals often 'sample' food for nutritious items, being reinforced by internal stimuli, but also react directly on chemical cues as fatty substances, sucrose, alcohol and certain minerals. Plants defend themselves from being eaten by mechanical and chemical means, e.g. alkaloids and phenolic substances. Animals switch between food types according to density and often concentrate in patches with high availability. Animals try to keep lean to be mobile and able to avoid predation.

INTRODUCTION

Wild animals differ more or less conspicuously from domestic animals and man with regard to availability of food; the food supply varies strongly between lean and bountiful times. Wild animals show many adaptations to avoid the lean times, by caching, migrations to better places, etc. but usually they cannot fully avoid the bottlenecks. This treatise will consider variations in diet in nature depending on seasonal and other environmental shifts and some adaptations to avoid starvation.

Diet selection varies tremendously between various animal groups. I will therefore start with a survey covering most of the spectrum. However, I will then restrict my treatment to animals fairly similar to humans with regard to digestive ability and general food supply. I have been working particularly with rodents and birds and I will pick up examples of species

J.B. Owen et al. (eds.), Animal Models – Disorders of Eating Behaviour and Body Composition, 69–82.

within these groups that may be of particular interest to human dietary disorders. I do not mean to provide any definite homologies, only to indicate some possible background to human problems.

THE FULL DIET SPECTRUM

A horse differs strongly from a wolf in feeding and diet items. However, the main problem is more with the wolf than with the horse. There is an old story about horses that were kept for mining on Bear Island between mainland Norway and Spitsbergen and maintained on seal flesh all over the winters; in summer they would find some of their common grass food. These horses were said both to survive and to be useful in the mines. True or not, animals commonly feeding on green vegetation are able to survive for a fairly extended period on more concentrated food as cereals or even meat. Instead, animals adapted to meat, or meat and cereals, will not survive any longer time on grass. And grass or green vegetation is generally in a much larger supply than meat or grain. So the horse will face a much wider choice than the wolf.

The main difference between the horse and the wolf lies in the different ability to break down and at least partly digest the walls of plant cells. These walls are made up by cellulose, hemicellulose and lignin and the digestive enzymes of animals can usually not attack such substances (Hungate 1966). However, the horse keeps in his large intestines a microflora, mainly consisting of bacteria, that has such a capability. Thus, the horse is acting in symbiosis with these microbes to overcome the protection of many plants against herbivores. The horse has even developed a particular system of enlarged colon and caecum to host his benevolent, and necessary, partners. There is another group of herbivores that has taken this adaptation still further, *viz*. the ruminants. The latter has a whole battery of stomachs hosting various fractions of a comprehensive microflora. Ruminants as cows also help this flora by repeated chewing of the food plants. The wolf has nothing in this way and can definitely not survive on grass. However, we know that domestic wolves, i.e. dogs, can be in part supported by ground cereals, e.g., as bread. Seeds as cereals do not contain any very thick cell walls and grinding and boiling will destroy cell wall compounds. Still, there are strict limitations to how much cereals a carnivore can consume and utilise.

It is thus possible to establish a gradient, or spectrum, in food composition for the animal world (Figure 1). At one end of this gradient are animals that are able to utilise the worst of roughage and at the other those that are strictly limited to meat. Moss is considered as particularly difficult to digest (Prins 1982) and few animals do eat moss. However, lemmings are an exception and they have evolved enormous caeca harbouring a very extensive microflora. Grass has thicker cell walls than most dicotyledonous herbs, as dandelion, clover and alfa-alfa, and fewer animals are able to use grass than thin-walled green plants. Large ruminants and, particularly, horses eat much grass while small ruminants

Food	Feeder
Moss	Bryovores
	Lemming
Twigs, leaves	Browsers
	Moose
Grass, forbs	Grazers, folivores
	Horse, field vole
Fungi, lichens	Fungivores
	Bank vole, flies
Fruit, berries	Frugivores
	Monkeys
Seeds, fruits	Granivores
	Wood mice, titmice
Meat	Carnivores
	Wolf, weasel

Figure 1: The food and feeder spectrum. Some typical species are indicated for the various food types.

and many folivorous rodents (e.g. bank and field voles) prefer the tender herbs (Demment & van Soest 1985). Other rodents, that usually are called granivorous (e.g. mice and rats) eat mainly seeds, or seed-type food. This is also the main food of many passerine birds while others are insectivorous, relying on more or less pure meat. However, insectivorous birds and bats are not able to digest chitineous structures as wings and body shields of the insects. Finally, some carnivorous mammals and birds as weasels and hawks are strictly restricted to red meat. All this variation in diet composition is reflected in the anatomy of the digestive system. Although large homeotherms (mammals and birds) are generally supposed to be able to digest rougher vegetation than small homeotherms, certain small mammals and birds, as the lemmings, have developed a very intricate digestive system in the colon and caecum to attack roughage (Foley & Cork 1992). This applies to both the hindgut morphology and physiology as well as to the ability to enlarge the intestine to some extent at energy crises. Digestibility of food biomass varies from less than 50 % in grass eaters to 90-95 % in pure meat eaters.

Few animals are limited to just one group of food but rather select, or have to select, from adjoining groups in this food gradient. Thus, bryovorous lemmings eat grass in summer when available, field voles eat delicate herbs in summer and grass in winter, and mice eat insects and other animal food in spring-summer and seeds in autumn winter. Restrictions to one particular food type, and particularly to one food species, is very rare, at least in temperate environments. Still, animal species are more or less specialised with regard to the plant and animal species eaten. The fox, a generalist carnivore, may eat hares, grouse eggs and young, roe deer fawns and rodents while a weasel is more or less restricted to the rodents.

Granivores and carnivores eat minute amounts of plants. These plant parts evidently do not supply any important portion of the energy, or gross nutrients, needed by these animals. The reason for this plant consumption is obscure; certain authors suggest it will help digestion and others that it supplies vitamins or minerals necessary in trace amounts. Vitamin K and some vitamins in the B group are synthesised by the gut flora and that may be one reason for minor plant consumption in non-herbivores.

Where to place humans in this gradient? No human being can survive on grass and several authors have suggested that man originated as a meat scavenger, and partly predator, on the African plains. Some extant aboriginal people as the Inuits live mainly on meat while agricultural people eat cereals, boiled vegetables and meat. Humans are probably most like rats and mice in having seeds and meat as original food but being able to extract energy and nutrients from tender vegetables as juicy roots, stems, etc. by grinding and/or boiling. Still, we are not able to digest heavy-celled vegetables as grass even after extensive treatment. We probably always have eaten minor amounts of tough plants for obtaining important trace compounds. To make the following overview relevant for understanding human feeding, I will particularly consider granivorous rodents and birds. However, I will put little emphasis on rats and house mice as they now live in fairly artificial environments. I will instead consider their wild relatives and particularly the wood mouse (*Apodemus sylvaticus*), perhaps a model species for human feeding! This species is common in western Europe but the more east-European yellow-necked mouse (*Apodemus flavicollis*) and the American deermice (*Peromyscus* spp) show similar feeding behaviour.

SEASONAL AND ANNUAL VARIATION IN DIET AND FEEDING

Wood mice living in deciduous woodland eat a lot of insects, e.g. lepidopteran caterpillars, in spring and early summer (Hansson 1985a, Figure 2A). Also adult butterflies or moths are taken, recognised by the wing scales at diet analyses of stomach contents or faeces. Chaete of earthworms are also often seen at such analyses and earthworms may make up a considerable part of the summer food. However, this animal food is very easily digested and the quantities consumed are difficult to estimate. Towards the end of the summer berries and small seeds appear in the diet. Green leaves make up 10 % or less. Autumn is a crucial season for such mouse populations. Many forest trees show great variation in seed setting and there is a certain rhythm in fructification, also affected by the summer temperature at bud formation (Pucek et al. 1993). Thus, every sixth to ninth year there will be copious amounts of forest seeds as acorn and beechmast and the trees are said to mast. However, there may be reasonable seed crops also other years while the years after mast seeding usually demonstrate nil seed production, obviously due to physiological exhaustion of the trees. So the mice will either eat almost only large tree seeds in autumn or continue with a mixed diet of animals, berries and fruit (as far as available) and small seeds from herbs in the field layer. Fungi may to some extent replace seeds. If there is a good supply of large seeds the animals will store them in well dispersed caches of various sizes. Thus, they will be available within the home range of a mouse all through the winter and early spring. When these seeds are lacking conditions turns difficult for the mice and survival declines; population densities are much larger in springs after good seed crops than after the poor ones. In the latter case the mice search for hibernating insects and other inactive invertebrates. Sometimes they even take vertebrates as birds in nests, or conspecifics. The protein and fat content of the stomach contents has been found to decline from spring-summer to autumn-winter, obviously in relation to a simultaneous decline in the consumption of animal food.

The wood mouse also demonstrates differences in diet depending on habitat. Individuals living in coniferous forests in central Europe consume a larger proportion of animal food and less seed than in the deciduous forests (Figure 2B). Spruce seeds that are shed in late winter were predominantly eaten in spring. Grass flowers were a food component specific to this habitat. Wood mice in agricultural land ate mainly cereal and weed seeds. Certain differences between population categories have been reported: Females consumed more green matter than males and very young free-roaming individuals ate fewer seeds and a wider spectrum of other food items. For granivorous birds this difference is aggravated:

Although the adult birds may eat seeds all year round they feed their nestlings with insects as the young require protein for growth. Females of both birds and mammal require more minerals, particularly calcium during breeding (Barclay 1994).

The varying food composition affects the size and functioning of the intestinal tract. The alimentary canal of pregnant and, especially, of lactating females exhibits increases in length and hypertrophy of the walls, also repeatedly in time at multiple litters. In particular the caecum

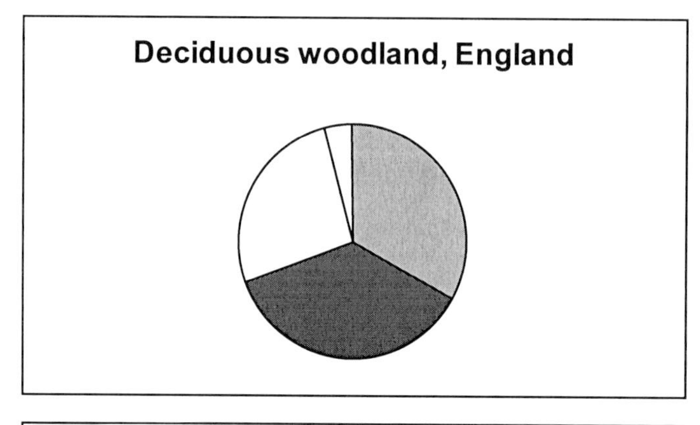

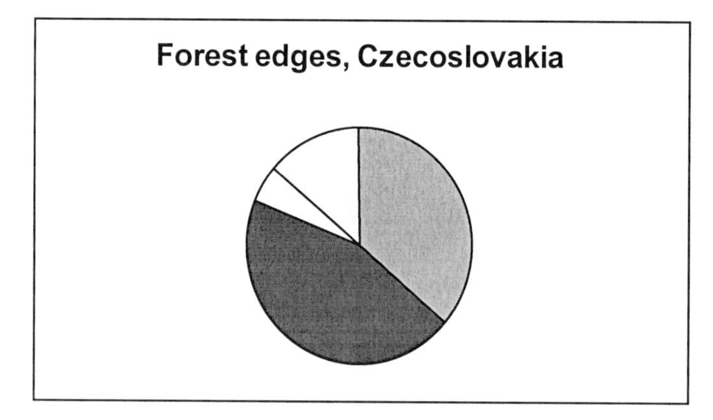

Figure 2A: Food composition of the wood mouse *Apodemus sylvaticus* in spring in three different habitats. The food items are clock-wise animal food, seeds and fruits, green leaves and various, including fungi. From Hansson 1985a.

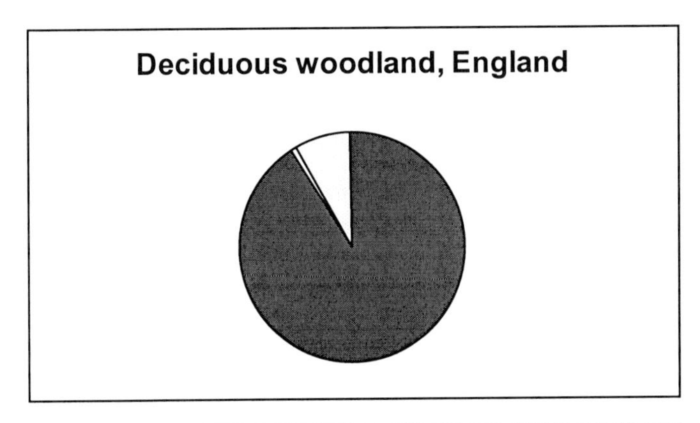

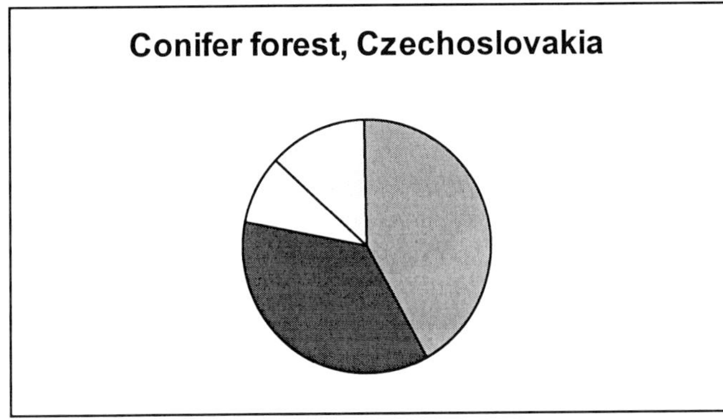

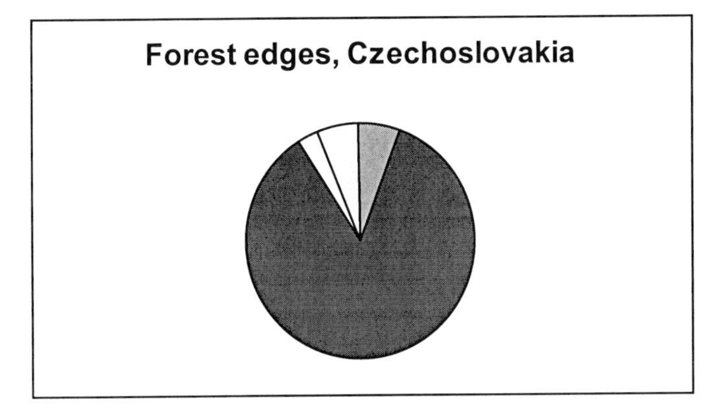

Figure 2B: Food composition of the wood mouse in autumn in three different habitats. Explanations as in Fig 2A. From Hansson 1985a.

becomes enlarged during lactation, which is very energy-demanding. For the bank vole the energy requirements have been found to increase with 24 % during pregnancy and with 92 % at lactation (Kaczmarski 1966). At experiments with various types of food the intestines remained shorter at pure concentrates of cereals. There are also differences between boreal and temperate environments in intestinal lengths in the mixed granivorous-folivorous bank voles, indicating different feeding regimes (Hansson 1985b). Clostridium bacteria that ferment cellulose under unaerobic conditions are particularly common during winter and early spring in the bank vole (Kunicki-Goldfinger and Kunicka-Goldfinger 1962).

Wood mice produce litters all through the summer period May-September. There is usually a non-breeding period, with undeveloped or regressed gonads, in October-April but winter breeding occurs after autumns that are particularly rich in seeds, e.g. after mast seeding in oaks and beech. Even at such rich food supply the animals do not become particularly fat. Indeed, fat individuals of this species are almost never seen in nature. In other species of granivorous rodents as the hazel mouse and its relatives that truly hibernate with lowered metabolism, the animals instead accumulate large quantities of adipose tissue in autumn before hibernation. These animals even get so fat that they were once eaten by the Romans as a delicacy. Neither do more folivorous but winter-active species as field voles accumulate discernible fat during any season. However, when brought into the laboratory and kept in small cages with food ad libitum all species of small rodents become very fat, evidently independent of exact diet.

Not only mice exhibit seasonal or annual food shortage that affects their diet. Large herbivores as deer and reindeer may show signs of starvation in areas without predators or when released on (small) islands (Klein 1968). These animals then often increase to very large numbers and high densities but have to change diet to less palatable plant species, particularly during winter-time. Certain species of small rodents (however, not the wood mouse) are known for their 'cyclic' population changes with peak densities every three-four years and following population 'crashes'. During peak winters folivorous rodents as field voles consume bark of coniferous seedlings and trees, a food item that is not eaten at other times and never touched at laboratory feeding with surplus of other food. This bark consumption can be so widespread that economically great damage is done to forestry.

CUES IN DIET SELECTION

Many factors are involved in the selection of distinct food items, plant or animal species or parts of species. Many animals are known to 'sample' vegetation for most palatable plant parts and then remain with this food type for a while, later on resuming sampling. Some insects attack preferentially physically stressed plants as the concentration of nitrogen is higher and the concentration of defensive substances is often lower. Thus, insect damage to plants is common after dry spells.

However, often the food is chosen directly from some distinct character. Colour and texture may sometimes be involved but most research has been directed towards olfactory cues. More general positive responses are common: Seeds buried in deep sand can e.g. be smelled and retrieved by various granivorous rodents (Jennings 1976). The ability of the rodents to sense the volatile seed components differs with environmental conditions; it is particularly poor during dry conditions (van der Wall 1998). However, a few specific chemical compounds can govern food choice as exemplified below.

Selection of various constituents has often been examined with wooden sticks impregnated with more or less pure substances. One of the better known substances in this respect is sucrose that attracts a lot of animal species. Evidently it advises the animals about ripe and useful fruit. Less known is perhaps the attraction of certain insects to ethyl alcohol; evidently suggesting the occurrence of decaying plant parts (Jonsell & Nordlander 1995). Experiments with this alcohol seem not to have been performed with mammals, except perhaps at cocktail parties. However, glycerol, being a more complex alcohol, is of high interest to rodents (Hansson 1973).

Fatty substances work also as attractants. This applies to common commercial table oils as well as to some of its constituents. Linoleic aid appeared most preferred by various rodent species, including the wood mouse, while e.g. tripalmitin, oleic acid and butyric acid were rejected (Hansson 1973). Juvenile rodents showed a greater interest in vegetable oil than the adults.

Another sort of substances that attract small rodents are minerals supplied as simple salts in wooden sticks. Similar effects have been achieved by fertilising tree seedlings and watching attack rates on the tree bark after one vegetation period. The most attractive mineral is evidently sodium while there have been preferences also for calcium (Hansson 1990a). Phosphorous has instead been rejected in some tests. The 'hunger for salt' (sodium chloride) is well known throughout most of the animal world, however particularly among herbivores (Denton 1980). Moose select aquatic plants to get salt, salt stones are provided for wildlife, cows look eagerly for all kinds of salty substances, including urine, reindeer lick roads for salt (however $CaCl_2$!) and deer and sheep massacre nestling birds (Bazely 1989). The massive consumption of experimental salt supplies is probably not only a superoptimal stimulation; small rodents consume disproportional amounts at cyclic peak densities (Hansson 1990a) and they may then be in nutritionally or, more generally, physiologically bad shape. The functional basis for this salt drive has been supposed to be a need of sodium in the detoxification process of plant secondary substances (Persson 1983). Still, sodium fulfils many import physiological processes, particularly in the nervous system.

There are other preferences that are not very easily understood. One of them is the heavy attack rate (gnawing) of the plastic covering of electric cables by rodents!

THE PLANTS FIGHT BACK: REPELLING SUBSTANCES

Plants usually do not like being eaten, some fruits and berries being exceptions. Thus, many plants have developed means to make them unpalatable. Being thorny or prickly is one mean, synthesising nasty or poisonous substances is another. With regard to chemical defence one usually separates between inherent and induced defence. The inherent substances are always in the plants, while induced substances are synthesised after incipient consumption or even just the presence of destructive herbivores. Induced defence is usually related to insect herbivores, which do not destroy an entire plant individual at once. It has been assumed to work also for mammalian herbivores but there is hardly any convincing evidence yet for these larger animals.

The inherent defence is thought to develop in different ways depending on the general supply of nutrients to a particular plant species (Bryant et al 1980). Species growing on very fertile soil synthesises metabolically 'expensive' compounds that are very poisonous or lethal to the herbivores. These compounds often contain nitrogen. Many alkoloids, sometimes used in human medicine or in drugs, are of this type. Such plants are said to have a qualitative defence. Other plants, growing on poor soils or in harsh climate, instead use surplus production of carbohydrates to synthesise phenols, condensed phenols as tannins, or terpenes. Such substances are usually carbon-based. The tanning tannins of oak leaves are one example. They work in the herbivores by reducing or preventing digestion of the food, often affecting the symbiotic bacteria needed for cleavage of long carbohydrate chains. Such a defence, varying in strength depending on external conditions and affecting digestion rather than sensitive metabolic pathways is said to be quantitative. It has also been argued that lignin, an important component of the plant cell walls, has developed as a protection against herbivory. Indeed, most herbivorous animals prefer young growing plant parts with relatively little lignin.

Most of this defence is found in green plant parts. However, both seeds and bulbs eaten by animals sometimes contain very poisonous compounds. Thus, two commercial poisons, that earlier were used against rats and mice (rodenticides), were extracted from bulbs (*Urginea maritima*) and nuts (*Strychnos nos vomica*), respectively. They have now usually been replaced by milder synthetic compound due to the great risk also evoked to humans by their use. Also animal food can be toxic; the Spanish fly (*Lytta vesicatoria*) is a case in point.

Some animals, perhaps particularly insects, have overcome the toxic effects by evolutionary adaptations in the digestive system and use these substances as stimuli for recognition of favourable food plants. They may even store them in their own body tissues as a defence against predators (Rothshild 1973).

Deterrent substances may not be sensed by an animal before the first consumption. Goats and sheep may sample a variety of foodstuffs, more or less digestible and even more or less toxic, and then show a post-ingestive feedback on the food and its consequences with regard to nutrition and health. Less suitable foodstuff will not be sampled or consumed again (Provenza 1995).

HOW AND WHERE TO FEED?

An animal wants to spend as little energy as possible in securing its food. Therefore it should select energy-rich food and patches with high food density when foraging. These general considerations have led to a lot of theorising on animals feeding behaviour (Stephens & Krebs 1989). Some conclusions of the optimal foraging theory are that the animals should prefer the most profitable prey (highest energy return per effort), be more selective when profitable food is common and disregard unprofitable food. The problem about how long to remain in a food patch has been explored according to the marginal value theorem. A main conclusion has been that an animal should remain in a profitable patch until its rate of intake drops to the same level as the average intake for all its habitat.

Within a certain range wood mice switch between food items, e.g. animal food and large seeds, according to abundance. Experiments with American deermice in a classic study (Holling 1959) have even demonstrated that the consumption in relation to food abundance (in theory called the functional response) follows well a sigmoid curve. There is first a low consumption at low abundance, then a rapid increase in consumption at small increments in abundance and finally an asymptotic consumption at further large increases in food abundance. Such foraging according to a sigmoid curve is interesting as it theoretically may keep prey abundance on an even level.

The general theory is indeed fulfilled for the wood mouse and similar small mammal species even with regard to spatial distribution of foraging. These animals are concentrating their foraging to patches with mass occurrence of defoliating lepidopteran caterpillars. This foraging behaviour may be important in the biological control of e.g. forest pest insects. However, the theory is now known to mainly apply to animals maximising energy intake. For animals that try to obtain an optimal mix of nutrients and sufficient amount of energy (applies particularly to grazers) a far more complicated theory has been developed (Belowski & Schmitz 1994) – and only tested to a very limited extent.

A practical problem when applying such theories is that many animals display neophobia when confronted with a new food type; they do not touch even very nutritious food until after several days, if at all. This is one reason why many rodenticides fail; rats particularly do not easily eat new food types that suddenly appear in their well-known immediate environment.

WHY ANIMALS KEEP LEAN

Most free-living animals are lean. They may be somewhat fatter in winter than in summer. Larger mammals have a relatively larger fat store than small mammals. Larger reserves ought to reduce the risk of starvation. The reason why animals do not accumulate large depots of fat has been an important research topic that, however, has been investigated mainly in birds. Special attention has been paid to differences between dominant and subordinate individuals in winter flocks.

Food resources show both seasonal and daily variation in abundance or availability. Metabolic requirements vary also over time, e.g. depending on environmental temperature. Accumulation of fat reserves imposes, however, a survival cost, particularly in reducing a prey species' agility for avoiding predators. Therefore, some intermediate fat level should be optimal. Subordinates have to wait to feed until the dominants are satisfied, alternatively subordinates have to rely on less abundant resources. Therefore, subordinates face more unpredictable food conditions than dominants. Subordinates should thus carry larger fat reserves, something that also has been observed in titmice (Rogers 1987, Clark & Ekman 1995) that can be both granivorous and insectivorous. Removal of dominant birds have led to less fat accumulation in the subordinates. In other cases food availability has limited the reserves in the subordinates.

Many birds have also a pronounced daily rhythm in body reserves. Large parts of them are built up from morning to evening. The bird senses the time of the day and environmental conditions, particularly temperature, and is regulating its intake to near-optimal night deposits. Again titmice as the great tit has demonstrated such foraging behaviour. Such restrictions in intake rates are particularly surprising when considering that fasting endurance appears to be a main way for northern animals to survive bad weather or temporary limitations in food access (Lindstedt & Boyce 1985). Winter-active small mammals carry fat reserves that will permit total starvation for only a day or two. Predation risk is evidently a very important factor to cope with at feeding.

Many mammals, particularly small mammals, show an annual rhythm in body weight. The body, including the skeleton, declines in size from autumn to winter, to increase again in late winter-early spring. The weight decline can be 10 % or more in late autumn (Hansson 1990b). Several adaptive reasons have been suggested, a more likely one being that total metabolism is lessened in the non-breeding winter state in order not to wear the body machinery.

WILD ANIMALS AS MODELS

It might be possible to understand several aspects of human feeding behaviour by analogies to wild animals. However, it is then very important to select suitable model animals as e.g. grazers and pure carnivores differ profoundly in food choice, digestion and nutrient requirements. Some suitable groups of species have been indicated in this short overview.

And a very short overview it is definitely, not being based on any deliberate, comparative studies of wild-living animals. However, it might be possible to design such studies on appropriate species with particular problems in mind in order to further our comprehension. I will end with a short list of such possible investigations:

1. How do growth, reproduction and aging affect food choice?
2. How is the immediate food choice affected by sudden changes in food supply? Are there important delay effects?
3. How do animals behave in difficult food situations? Eating inferior food, moving away or accepting partial starvation?
4. Do animals respond differently to food resources that are elevated in unexpected and predictable ways?
4. How do needs for micro-nutrients affect feeding?

Our knowledge about the evolutionary background to human feeding behaviour would certainly increase at joint research among physicians, veterinarians and animal ecologists!

References

Barclay RMR. Constraints on reproduction by flying vertebrates: energy and calcium. *Am Nat* 1994; 144: 1021-1031.

Bazely D. R. Carnivorous herbivores: Mineral nutrition and the balanced diet. *TREE* 1989; 4: 155-156.

Belovsky GE Schmitz OJ. Plant defence and optimal foraging by mammalian herbivores. *J Mammal* 1994; 75: 816-832.

Bryant JP Capin III FS Klein JR. Carbon/nutrient balance of boreal plants in relation to vertebrate harbivory. *Oikos* 1993; ?: 357-368. ??

Clark CW Ekman J. Dominant and subordinate fattening strategies: a dynamic game. *Oikos* 1995; 72: 205-212.

Demment MW van Soest PJ. A nutritional explanation for body-size patterns of ruminant and nonruminant herbivores. *Am Nat* 1985; 125: 641-672.

Denton D. *The hunger for salt.* Berlin: Springer. 1984.

Foley WJ, Cork SJ. Use of fibrous diets by small herbivores: how far can the rules be bent? *TREE* 1992; 7: 159-162.

Hansson L. Fatty substances as attractants for *Microtus agrestis* and other small rodents. *Oikos* 1973; 24: 417-421.

Hansson L. The food of bank voles, wood mice and yellow-necked mice. *Symp Zool Soc London* 1985a; 55: 141-168.

Hansson L. Geographic differences in bank voles *Clethrionomys glareolus* in relation to ecogeographical rules and possible demographic and nutritive strategies. *Ann Zool Fennici* 1985b; 22: 319-328.

Hansson L. Mineral selection in microtine populations. *Oikos* 1990a; 59: 213-224.

Hansson L. Ultimate factors in the winter weight depression of small mammals. *Mammalia* 1990b; 54: 397-404.
Holling CS. The components of predation as revealed by a study of small-mammal predation of the European pine sawfly. *Can Entomol* 1959; 91: 293-320.
Hungate RE. *The rumen and its microbes*. New York: Academic Press. 1966.
Jennings T J. Seed detection by the wood mouse, *Apodemus sylvaticus*. *Oikos* 1986; 27: 174-177.
Jonsell M, Nordlander G. Field attraction of Coleoptera to odours of the wood-decaying polypores *Fomitopsis pinicola* and *Fomes fomentarius*. *Ann Zool Fennici* 1995; 32: 391-402.
Klein DR. The introduction, increase and crash of reindeer on St. Matthew Island. *J Wildl Manage* 1968; 32: 350-367.
Kaczmarski F. Bioenergetics of pregnancy and lactation in the bank vole. *Acta Theriol* 1966; 11: 409-417.
Kunicki-Goldfinger W, Kunicka-Goldfinger W. Intestinal microflora of wild animals – *Sorex araneus araneus* L. and *Clethrionomys glareolus glareolus* Schreb. *Acta Microbiol Pol* 1962; 11: 43-110.
Lindstedt SL, Boyce MS. Seasonality, fasting endurance, and body size in mammals. *Am Nat* 1985; 125: 873-78.
Persson A. Digestibility and retention of food components in caged mountain hares *Lepus tidmidus* during the winter. *Holarct Ecol* 1983; 6: 395-403.
Prins HHT. Why are mosses eaten in cold environments only? *Oikos* 1982; 38: 374-380.
Provenza FD. Postingestive feedback as an elementary determinant of food preference and intake in ruminants. *J Range Manage* 1995, 48: 2-17.
Pucek Z, Jedrzejewski, W, Jedrzejewska B, Pucek M. Rodent population dynamics in a primeval deciduous forest (Bialowieza National Park) in relation to weather, seed crop, and predation. *Acta Theriol* 1993; 38: 199-232.
Rothschild M. "Secondary plant substances and warning colouration in insects" In *Insect/plant relationships*. H. F. van Enden, ed. London: Blackwell. 1973.
Rogers CM. Predation risk and fasting capacity: do wintering birds maintain optimal body mass? *Ecology* 1987; 68: 1051-1061.
Stephens DW, Krebs JR. *Foraging theory*. Princeton: Princeton University Press. 1986.
van der Wall S. Foraging success of granivorous rodents: effects of variation in seed and soil water on olfaction. *Ecology* 1998; 79: 233-241.

Chapter 5

THE ANIMAL WITHIN: LESSONS FROM THE FEEDING BEHAVIOUR OF FARM ANIMALS

Ilias Kyriazakis
Animal Nutrition and Health Department, Scottish Agricultural College, King's Buildings, West Mains Road, Edinburgh EH9 3JG

ABSTRACT Modern genotypes of farm animals have been selected intensively for particular traits and may be given access to foods that would not have been part of their evolutionary history. In this chapter the question is raised of whether the feeding behaviour of such changed genotypes can be understood within an evolutionary framework. The literature reviewed suggests that it can and that when the feeding behaviour of farm animals is inconsistent with this there are two possible kinds of explanation. The first is it may arise from the particular conditions of the experiment, which may confront the animal with an unsolvable problem. The second is that it is a reflection of the selection for some traits other than those that lead to fit and healthy reproductive adults. It is concluded that a modified evolutionary framework can account for the feeding behaviour of farm animals. Human subjects in affluent Western societies can also be seen as genotypes that have been subjected to 'artificial selection' and are given access to 'artificial foods'. The many parallels in the underlying control of feeding behaviour between farm animals and humans, suggest that it is reasonable to account for human feeding behaviour by the same framework used to account for farm animal behaviour, at least in part. There may well be heuristic value arising from studies of farm animals that is relevant to human feeding behaviour. However, as influences of other kinds (eg societal, self image, human purpose) impinge uniquely on the control of feeding behaviour in humans, it is unlikely that there will be a one-to-one correspondence in the controls and outcomes of feeding behaviour between the two.

INTRODUCTION

In the ecological literature it is implicitly assumed that all behaviours have evolved in a way such that animals will make optimal use of their natural environment; this is consistent with the idea of evolutionary fitness. In relation to feeding behaviour the assumption is that the amount

J.B. Owen et al. (eds.), Animal Models – Disorders of Eating Behaviour and Body Composition, 83–96.

eaten, and the diet selected from different foods, will be consistent with adaptive behaviour and, hence, those which are best suited to the needs of the animal. Such behaviour would be expected to enable animals to grow, reproduce and rear their offspring, and thereby ensure the survival of their genes. A question that then arises is whether the same principles are expected to apply to laboratory and farm animals. In relation to the latter, the expectation is indeed that farm animals which are kept in extensive environments, such as those grazing pastures, and have been subjected to a modest degree of artificial selection will also be able to show a feeding behaviour consistent with evolutionary fitness (Hutchings *et al.*, 1999). These environments are not dissimilar to those in which animals have previously evolved.

On the other hand, the majority of farm animals have now been selected for traits such as fast growth, high or low lean to fat ratio in their bodies and high yields of milk or eggs to an unusual extent. In addition to the resulting spectacular genetic changes (Emmans and Kyriazakis, 2000), animals are now kept in highly controlled environments and are frequently offered access to foods that in most cases are very different from the ones they have evolved upon. Examples would be the highly concentrated foods offered to herbivore sheep and cattle and the foods of high protein content fed to young chickens and turkeys. The question then is whether one should expect that, under such conditions, animals would continue to show feeding behaviour which is consistent with the idea of evolutionary fitness. Consideration of this question constitutes the subject of this chapter. The view has already been put forward that aspects of feeding behaviour in farm animals may no longer have their original function and thus will be difficult to understand (Moss, 1991). There are certainly examples of feeding behaviour of laboratory and farm animals, which appear not to be consistent with the idea of adaptive behaviour and these present a problem. An apparent alternative to seeing the feeding behaviour of farm animals in an evolutionary context is that of reducing the behaviour to its elements. The daily rate of food intake maybe expressed either as the product of meal size and meal frequency or, in the grazing animal, as the product of bite size, bite rate and grazing time. This apparent alternative approach turns out, on closer examination, not to be satisfactory (Emmans and Kyriazakis, 2001).

The issue of whether feeding behaviour in an 'unnatural' environment can still be explained by the principles of evolutionary fitness, is pertinent to the study of human feeding behaviour, and hence the subject of this book. The environment of people living in affluent Western societies can be seen as being 'artificial'. Food availability is constant and abundant, and the processing of foods can greatly disguise their nutritional value. The last part of this chapter will, therefore, relate to the main theme: on whether we can advance our understanding of human feeding behaviour by considering the principle of evolutionary fitness. In other words, is it appropriate to use laboratory and other animals as models for the study of human feeding behaviour? The alternative is that human feeding behaviour should be viewed completely independently and *sui generis*. It will be shown that animal studies can lead to ideas that have heuristic value to the study of human feeding behaviour, but that there is more to humans than 'the animal within'.

IS THE FEEDING BEHAVIOUR OF FARM ANIMALS ADAPTIVE?

This chapter concentrates mainly on the feeding behaviour of non-ruminant farm animals, ie pigs and poultry, as they can be seen as more appropriate experimental models that can be related to human feeding behaviour than can ruminant animals. In addition, the feeding behaviour of the latter appears to be heavily influenced by issues that arise from their complex and symbiotic rumen environment (Kyriazakis *et al*., 1999). In the first instance, the literature of feeding behaviour of non-ruminant animals contains as many examples of inconsistent behaviour as it does of feeding behaviours consistent with the concept of evolutionary fitness. This has led to two deeply divided camps, which either oppose or support the view that farm animals are able to control their food intake and select a diet that is consistent with their needs. A critical review of the literature reveals that observations inconsistent with adaptive feeding behaviour fall into two categories: either genuine failures of the animals to show adaptive feeding behaviour, in the manner we understand it at least, or failures that arise from the conditions of the experimental set-up. I shall return to the former later, but for the time being I will address the issue of how experimental conditions can lead falsely to the conclusion that farm animal feeding behaviour is not adaptive. Such conditions relate to three issues: (i) the prior knowledge of the animal of the food or foods given access to; (ii) the duration of the experiment (timescale of feeding behaviour); and (iii) the composition of the available foods.

Prior Knowledge of the Foods

It is now accepted that animals need to have some prior knowledge of the available foods if they are to show feeding behaviour that enables them to meet their needs. With the exception of certain 'hard-wired' behaviours, such as an innate preference for sweet foods and an avoidance of bitter ones (Leathwood and Ashley, 1983) and the possible existence of some, but not many, 'adaptive' specific appetites for nutrients such as sodium (Denton, 1982), all other components of feeding behaviour appear to need to be learned. This is consistent with the view that by allowing learning the animal is able to be more effective in adapting to the possibly complex temporal and spatial changes in its feeding environment. This is to say that, whilst the feeding environment on the farm can be constant and predictable, animals have evolved in an environment that had a high degree of variability and uncertainty.

In nature, many of the components of feeding behaviour are learned through associations and importantly they arise within a social context. Young animals that are expected to have a limited knowledge of their feeding environment, will be expected to acquire information about their feeding behaviour from their mothers and other conspecifics (Provenza, 1995). In many cases, however, young farm animals are abruptly separated from their mothers and are grouped together with other 'naive', same age individuals. Young pigs are weaned abruptly onto solid food together with other similar aged animals that also have no experience of solid food. Calves are removed from their dairy cow mothers immediately after birth and have no opportunity at all to learn from them. In the extreme case young chicks hatch having not even been incubated by their natural mothers, but in a machine. It is perhaps then not surprising, initially at least, that animals fail to recognise and consume foods that would be expected to have beneficial effects. In many cases they are able to show a feeding behaviour consistent with their needs, following an initial period of learning, during which they are eventually able to become familiar with the available foods and form associations between the properties of the foods and the post-ingestive consequences that they have (Kyriazakis *et al.*, 1991; 1999).

The speed of such learning is likely to depend on the current state of the animal in relation to its current needs. The greater the departure from the appropriate state, the greater will be the reinforcing properties of the appropriate food and the faster the learning. For example, animals that have previously been deprived of a nutrient can be expected to select very rapidly the supplemented food or appropriate diet, since under these circumstances the reinforcing properties of feeding can act very rapidly indeed (Capaldi *et al.*, 1991; see also below).

Timescales of Feeding Behaviour

Although the smallest unit of feeding behaviour in farm or laboratory animals, where feeding behaviour is recorded, is that of a visit to a feeder, it has been suggested that a more appropriate and meaningful unit to express short-term feeding behaviour would be that of a meal (Tolkamp and Kyriazakis, 1999). The latter relates more closely to biological principles, such as those of satiety, whereas a visit is subject to random influences, such competition at the feeder or external disturbances. The question then is whether we should expect that feeding behaviour, in terms of amounts eaten and diet selected, would be controlled within the timescale of a meal. This question has occupied the time and effort of many researchers and the consensus is that whilst feeding behaviour is very flexible, to the point of being seen as 'chaotic' in the short-term, it is much more orderly and controlled within the longer term (Collier and Johnson, 1990; Kyriazakis *et al.*, 1999; Emmans and Kyriazakis, 2001). It can be argued, therefore, that some of the failures of animals to show feeding behaviour consistent with evolutionary fitness is merely a reflection of the fact that the timescale of the observations of feeding behaviour has been too short.

The above suggestion, however, seems to contradict examples where feeding behaviour appears to be controlled within a very short timescale indeed. For example, it has been reported that egg laying hens select a Ca-supplemented food following a two-day deprivation caused by feeding on a very low Ca diet, within 15 min from being offered a choice between a Ca-supplemented and an unsupplemented food (Hughes, 1979). For this reason, Kyriazakis (1997) and Kyriazakis *et al.* (1999) have suggested that the appropriate timescale within which feeding behaviour is controlled would be related to the animal's internal state and specifically in the change created to it. This change could be either as a result of physiological change or as a consequence of previous feeding. In relation to the above example, this suggestion seems to account not only for the rapid response of the deprived hens, but also for the diet selection for Ca of non-deprived hens and pullets. For non-deprived hens the Ca preference developed more gradually, ie over a period of hours rather than minutes. Calcium deprived pullets, which have presumably significantly lower Ca requirements than laying birds, took two to three days before they showed a strong preference for a Ca-supplemented food (Hughes and Wood-Gush, 1971).

It is concluded, therefore, that if appropriate feeding behaviour is to be observed, then the timescale of the observation should be of sufficient duration. This, however, raises the question of how long the observations need to last ? This problem is overcome by the suggestion that the timescale will depend on how much change has been created in the animal's internal state, which has to be rectified by subsequent feeding behaviour. The amount of change will also be related to the size of the animal, specifically its metabolic size (Kyriazakis *et al.*, 1999) and level of production or output. Real or actual time will be longer in physiological or metabolic terms for a smaller sized animal or an animal of the same size that has a higher rate of output.

Characteristics of Available Foods

In the simplest of cases, animals can be given access to a single food that is balanced in all its nutritional dimensions. The animal can then be seen as having the problem of needing to control intake in a single dimension, for example that of energy, in order to meet its requirements in all other nutrients. In the situation where the animal is given a choice between two foods, the simplest case is when the two foods differ also in a single nutritional dimension. In our laboratory, we have adopted this methodology when investigating the feeding behaviour of animals given a choice between two foods that differ in their content of protein (eg Kyriazakis and Emmans, 1991; Kyriazakis *et al.*, 1993; 1999). These experiments were done partly with the view of arriving at measurements of the protein needs of animals at different physiological stages. The outcome of the experiments on pigs has been consistent with the view that they are able to select a diet which is a reflection of their protein needs, whilst at the same time they avoid an excess protein intake. In addition diet changes over time in a way that matches the presumed change in requirement (Kyriazakis *et al.*, 1999).

The composition of a single food can be such that it is imbalanced in more than the one aspect, such as its protein to energy ratio. The two or more foods offered as a choice can, and in practice usually will, differ in more than a single nutritional dimension. The complexity of available foods is often inevitable and closer to the situation the animals might face in their natural environments. The more complex the composition of the food(s), the more difficult it would be for the animal to solve the problem of meeting its needs through its feeding behaviour and for us to interpret its outcome. It is possible that some of the apparent failures of animals to control, in the expected way, their intake and diet selection can be attributed to the complexity of the feeding environment. In order to deal with such situations we have been using a trade-off framework that allows us to interpret feeding behaviour. Where one outcome, in terms of amounts of food eaten and diet selected, has both higher benefits and higher costs than the other, then the animal has to make some trade-off between the increased benefit and the increased cost in order to come to a decision.

Conventionally, the trade-off framework has been used to account for the feeding behaviour of animals given access to a food or foods that differ in their nutrient and toxin concentrations (Belovsky and Schmitz, 1994). Other examples in which the animal may be seen as having to make a simple trade-off are when it is given access to a single food that has a protein:energy ratio lower than that which is needed. One expectation is that the animal will increase its intake of such a food in order to try to meet its protein needs. This will have the consequence of extra energy being consumed, which could be associated with costs such as those of dissipating extra heat to the environment or coping with the excess bulk intake. The intake of the animal will then be the outcome of the weighting the animal gives to the relevant benefits and costs. It is in this way that we now view the outcome of the food intake of young pigs given access to foods of different protein content (Kyriazakis *et al.*, 1991).

The idea of trade-offs is a useful framework within which the feeding behaviour of animals can be viewed and interpreted. It necessarily raises the question of what is the 'currency' that the animal uses when it weighs the costs and benefits of its feeding decisions. Expressing these in a common currency, or on a common scale, so that feeding behaviour can be interpreted, is never straightforward. The approach also has inherent in it the risk that different, previously unthought of, trade-offs can be invoked to account for any unexpected feeding behaviour of the animals. This can lead to a theory where feeding behaviour, whatever it may be, can always be claimed to be consistent with evolutionary fitness. However, when used correctly the approach can be useful in interpreting the feeding behaviour of animals in complex situations. Here it should be sufficient to point towards the risks associated with it.

PARADOXICAL FEEDING BEHAVIOUR: FARM ANIMAL BEHAVIOUR INCONSISTENT WITH EVOLUTIONARY FITNESS

As mentioned previously, there are instances where the feeding behaviour of farm animals appears to be genuinely inconsistent with the idea of evolutionary fitness. These cases cannot be accounted for by the conditions of the experimental set-up and hence the nature of the problem needed to be solved by the animal. Here, I will use two examples of such 'inconsistent' feeding behaviour and discuss whether it would be fair to conclude that feeding behaviour in farm animals may no longer have its original functions. The two examples are by no means unique in the literature of feeding behaviour of farm animals.

Feeding Behaviour Leads to Unfit Reproducing Adults

There are examples where the feeding behaviour of growing animals offered access to a high quality food and kept under 'ideal' environmental conditions, leads to unfit reproducing adults. These are the cases of chickens selected for meat production (broilers) and sows of modern genotypes. Broiler breeders that are given *ad libitum* access to a food during rearing have a very low reproductive capacity (Table 1). Their reproductive capacity is greatly increased by giving them much less food during rearing. Unsurprisingly, it is still below that of chickens that have been selected specifically for their laying capacity (Hocking and McCormack, 1995). Growing gilts selected against body fatness, arrive at a reproductive stage with such a low body fatness that they are unable to sustain their reproductive ability beyond their first parity ('thin sow syndrome', Whittemore, 1998). They frequently have to be culled much earlier than their 'unimproved' counterparts.

In both cases, however, the feeding behaviour of the animals and its consequences appear to be the outcome of the genetic selection imposed upon them. Both chickens and pigs have been selected for some traits other than for a fit and healthy reproductive adult, which would have been the outcome of natural selection. Chickens have been selected for their meat yields at an age before reproduction and pigs have been selected against fatness at slaughter (Emmans and Kyriazakis, 2000, 2001). Their subsequent feeding behaviour is merely a reflection of these past selection goals. Despite the spectacular consequences of artificial genetic selection, the effects on feeding behaviour within an evolutionary context appear to be small (Siegel, 1993). In the case of the chickens, their reproductive ability and hence 'fitness' within the context of their system of reproduction, will now depend on their method of feeding and hence entirely on their keepers.

Table 1. The effect of feeding level during rearing (ad libitum or restricted) on the reproductive performance, up to 45 weeks of age, in two strains of hens: a 'meaty' type (broiler) and an 'egg-laying' type (layer). The number of 'abnormal' eggs laid includes eggs with cracked or soft shells (modified from Hocking and McCormack, 1995).

	Feeding methods		*No of eggs per 100 hen days*		
Strain	*Rearing*	*Egg laying*	*Produced*	*'Abnormal'*	*Hatchable*
Broiler	*Ad libitum*	*Ad libitum*	66	50	16
	Restricted	*Ad libitum*	75	20	55
Layer	*Ad libitum*	*Ad libitum*	93	11	82
	Restricted	*Ad libitum*	94	12	82

Parasitic disease leads to an increased nutrient demand and a decrease in food intake

Macro- and micro-parasites impose an increased metabolic and nutritional demand on their hosts, which is most frequently manifested by impaired growth, productivity and reproductive ability. The impairment is due to an increased loss of protein, damage of tissue (such as damage of gut tissue and mucoprotein secretion in gastrointestinal parasitism (Bown *et al.*, 1991) or erythrocyte damage in protozoan infections (van Dam, 1996)) and the prevalence of fever, which leads to increased heat production and muscle proteolysis (Thompson, 1990). Under these circumstances it would be expected that there would be some increase in the parasitised animal's food intake to enable it to overcome the taxing consequences of parasitism. Paradoxically, however, what is observed is a reduction in food intake, henceforth called anorexia, of the order of 30-60% (Kyriazakis *et al.*, 1998); this anorexia is the major factor contributing to the reduced performance of the parasitised animal.

In the first instance, such a paradoxical feeding behaviour appears to be contrary to the view of the adaptive nature of feeding. However, evidence suggests that anorexia could actually be viewed as a disease-coping strategy, which confers certain advantages to the parasitised animal (Kyriazakis *et al.*, 1998). This is consistent with the view that other phenomena that occur during disease, such as fever, have a similar adaptive nature. This suggestion can be tested when a parasitised animal that shows a degree of anorexia is fed by some means so that its intake is similar to levels seen prior to the onset of the infection. The experiments of Murray and Murray (1979), summarised in Table 2, have addressed this in mice infected with the bacterium *Listeria monocytogenes*; food intake was sustained through intragastric tubbing and there was a 'control' treatment to account for the effects of intubation *per se*. This experiment is not the only example where the issue has been addressed, but is perhaps the most elegant one (for a review see Kyriazakis *et al.*, 1998). Force-feeding of infected mice to a 'normal' energy intake (ie that of their uninfected controls) led to an increased mortality, and shortened survival time, compared with infected mice that were allowed to become anorexic. This has led the authors to conclude that anorexia during parasitic diseases is indeed a coping mechanism that confers advantages. Kyriazakis *et al.* (1998) summarised what these advantages might be; they arise from the fact that many nutrients have immunotoxic effects and can inhibit the development of an effective immune response during the development of the disease. Anorexia, therefore, during parasitic diseases plays the role of promoting an effective immune response in the host. The other hypotheses that have been put forward to account for further potentially beneficial consequences of anorexia have been summarised by Kyriazakis *et al.* (1998). They are consistent with experimental evidence, but importantly, they support the view that anorexia is a disease-coping strategy, part of the mechanism of recognition of parasite invasion by the immune system, which leads to a modification of the host's feeding behaviour.

Table 2. The food intake (as a percentage of uninfected control), mortality rate and survival time (days taken to die from the onset of infection) of mice infected with *Listeria monocytogenes* and their respective controls, whilst on different feeding regimes (data from Murray and Murray, 1979).

Group	n	% intake of controls (group 1) ($\bar{x}$ ± se)	Mortality (%)	Mean survival time (days ± se)
(1) Uninfected, *ad libitum* (controls)	30	-	0	-
(2) Uninfected, tube feeding + *ad lib*	10	103 ± 2.7	0	-
(3) Infected, ad *libitum*	30	58 ± 6.7	43	8.7 ± 1.5
(4) Infected, *ad libitum* + tube feeding	30	100	93	3.9 ± 0.9

Mice in groups 1 and 3 underwent intragastric intubation. Mice in group 2 were given approximately 40% of their intake through an intragastric tube. Mice in group 4 had their intake made equivalent to group 1 through intragastric feeding.

The above example, of anorexia during parasitic diseases, illustrates the point that feeding behaviour, initially expected not to have a purpose, actually has an underlying evolutionary cause. It is consistent with the view that changes in farm animals have had only a minor effect on adaptive behaviour (Siegel, 1993).

A FRAMEWORK FOR THE FEEDING BEHAVIOUR OF FARM ANIMALS

The evidence presented above suggests that the feeding behaviour of farm animals can still be understood from the same behavioural programmes and goals that are considered to have applied or apply to their ancestors and wild counterpart species. That is to say that animals artificially selected for traits which do not confer any evolutionary advantages show a feeding behaviour in an evolutionary context.

Because the idea of evolutionary fitness cannot readily be applied to farm animals, Emmans and Kyriazakis (1995) have suggested that it might be better to consider that all animals, including farm animals, have current output goals which they seek to achieve through their feeding behaviour. The purpose of the suggestion was to find a way of applying a general evolutionary framework to the specific problem of feeding behaviour of species of farm animals. Proposed current output goals include their growth rate, the rate of milk and egg production and the level of fatness. The consequences of this suggestion are, in most cases, identical to the framework that has fitness as a goal. A sow, that aims to achieve its goal of milk production, or a chicken that aims to achieve its goal of egg production, both aim at maximising their 'fitness' within the context of the 'artificial' selection imposed by humans. The attraction of the suggestion lies in the quantifiable nature of the goal and its generality. The modified framework has been applied with success in accounting for the feeding behaviour of farm animals (Emmans and Kyriazakis, 2001). It is within this framework that the feeding behaviours of human subjects can be viewed.

THE ANIMAL WITHIN: DOES HUMAN FEEDING BEHAVIOUR HAVE AN EVOLUTIONARY BASIS?

Evidence has been provided above, it is hoped convincingly, that the feeding behaviour of modern or artificially selected genotypes of pigs and poultry is consistent with the idea of evolutionary fitness. This is seen even when they are given access to artificial feeds, in a form and composition that has not been previously encountered in their natural history. Advances in medicine, science and engineering have allowed the survival of human subjects that have not been under the pressure of natural selection, and hence they too can be seen partly as 'artificial genotypes'. In addition, human foods, in Western societies at least, are of an extremely high quality, highly processed and are subject to little uncertainty. These attributes contrast greatly with the food qualities available to early hominids that were rarely exposed to food items of excellent quality, that could be obtained only during limited periods in the year and after a considerable effort (Foley, 1987; Mela and Rogers, 1998). There are thus great parallels between human subjects and farm animals in their current feeding environment, and the question whether human behaviour should also be expected to have an evolutionary basis does not seem to be an inappropriate one. It is best addressed by considering first the similarities and then differences between the two different animal classes, in that respect.

Similarities in Feeding Behaviour of Farm Animals and Humans

There are many similarities in the underlying controls of and parallels in the feeding behaviour of humans and farm animals. (i) The physiological apparatus that underlies feeding behaviour is the same in both. It is now widely accepted that human overeating and obesity has an inherent component, as does differences in fatness between different strains of farm animals. Certain inherited eating disorders, such as the Prader-Willi syndrome, appear to arise from the same physiology that leads to obesity in other animals (Trayhurn, 1997). (ii) Both humans and farm animals use multiple sensory cues to learn about the consequences of ingesting certain foods, in the manner discussed above. Such associations lead to preferences for beneficial foods and aversions for detrimental ones (Bernstein and Borson, 1986 for human subjects; Provenza, 1995 for farm animals). (iii) Human feeding behaviour is sensitive to the same changes in the internal and external environments, as is the case with animal feeding behaviour. The increases in energy intake seen in humans during pregnancy, exposure to cold or exercise, for example, are identical to the changes seen in the feeding behaviour of animals under similar circumstances (Tremblay and Doucet, 1999). Feeding behaviour is driven by current output in both animal classes. (iv) Food intake is regulated in humans in an entirely satisfactory manner in the long-term, when no particular conscious effort is involved (Taggart, 1962), although it can be subject to short-term fluctuations. This is similar to the weight equilibrium maintained for a long time by non-reproducing, mature farm animals. The expectation, therefore, would be that were the human to exhibit a feeding behaviour that was meant to meet the needs of its current state *only*, ie the animal within, then this would be consistent with evolutionary fitness.

Differences in the Feeding Behaviour of Humans and Farm Animals

These arise from possible differences in the goals of feeding behaviour between humans and farm animals, and the considerable influence that societal factors and efforts of will have on human feeding behaviour. If an animal is trying to achieve its needs as these relate to evolutionary fitness, then by definition this is its goal. There can be several such goals, and it was alluded to above that farm animals, selected against body fatness, exhibit a feeding behaviour that allows this body fatness at a particular degree of maturity to be attained (Kyriazakis *et al.*, 1993). In fact, if animals are made fatter in relation to this 'desired' fatness through feeding on an imbalanced food, they return to the desired level of fatness by choosing a diet that allows them to do so when given an appropriate choice (Kyriazakis and Emmans, 1991). In other words, they show 'compensatory thinning'; how quickly this is achieved depends upon the composition of the foods offered as a choice and the environment. Thus a desired level of fatness is vigorously defended against perturbation in both directions. The signal for such control may involve the leptin system (Trayhurn, 1998).

Under similar conditions of food availability, the suggestion is that human subjects tend to become obese. Is this because humans also try to achieve a genetically predetermined level of fatness which, however, is set at a higher level? The view has indeed been put forward that human evolution has selected for our physiology and behaviour to favour nutrient over-consumption rather than under-consumption (Mela and Rogers, 1997). This is because human feeding behaviour has evolved within a context of uncertainty, where nutrient sources have fluctuated widely in space and time. Such environments are still prevalent in developing parts of the world. Evolution, therefore, would have been expected to favour individuals that were able to accumulate body stores in situations of food abundance, in order to survive and reproduce when food was scarce. The hypothesis that obesity is a consequence of such feeding behaviour is an attractive proposition, but subject to considerable debate.

Lastly, human feeding behaviour is considerably influenced by societal factors and is subject to cortical control through possible efforts of will. In a highly encephalised species, it is hardly surprising that feeding behaviour is subject to such control. Issues may arise from the way our bodies look to us, or are believed by us to be perceived by others. Such effects appear to exert a considerable influence, and in some instances a greater one, than the physiological controls that underlie feeding behaviour. Other dramatic examples of the effects of human will on feeding behaviour are humans refusing to eat as an act of political defiance, or forcing themselves to eat to become grossly enlarged, as in the case of Sumo wrestlers (Nishizawa, 1976). Thus, the current view is that human feeding behaviour is the outcome of a balancing act between achieving physiological goals and the moderating influence of cortical control (Rozin, 1996). There does not appear to be any example in farm animals of such influence. It is therefore, in this respect, that human and animal feeding behaviour fundamentally differ, and where there would be little benefit in pretending that the two can be related.

CONCLUSIONS

It is concluded that the understanding of feeding behaviour of farm animals, including those that have been subjected to intensive selection for particular traits, can be fruitfully approached using a framework that arises from the idea of evolutionary fitness. That is to say that animals, such as current strains pigs and poultry, artificially selected for traits which would not confer any evolutionary advantage in a natural environment, and given access to foods which are often highly artificial, show feeding behaviour that can be accounted for using an evolutionary framework (Emmans and Kyriazakis, 2001). Human subjects, at least in Western societies, can also be seen as genotypes that have been subjected to 'artificial selection' and are given access to 'artificial foods'. As there are many similarities in the underlying control of feeding behaviour in non-ruminant farm animals (especially pigs) and humans, we are led to expect that human feeding behaviour could be explained, at least to a certain extent, by the same principles of evolutionary fitness. It is for this reason that studies on farm animals may have heuristic value to human feeding behaviour. However, in humans other dimensions, such as those related to societal influences and efforts of will, are relevant to the control of feeding behaviour. Therefore, it is unlikely that there will be a one-to-one correspondence in the controls and outcomes of feeding behaviour between humans and farm animals.

Acknowledgements. This work was in part supported by the Scottish Executive, Rural Affairs Department. I am grateful to Bert Tolkamp for allowing me to plagiarise the title of one of his papers and to Gerry Emmans for extensive comments on the manuscript.

REFERENCES

Beolvsky GE, Schmitz OJ. Plant defences and optimal foraging by mammalian herbivores. *J Mammalogy* 1994; 75: 816-832.

Bernstein IL, Borson S. Learned food aversion: a component of anorexia syndromes. *Psychol Rev* 1986; 93: 462-472.

Bown MD, Poppi DP and Sykes AR The effect of post-ruminal infusion of protein or energy on the pathophysiology of *Trichostrongylus colubriformis* infection and body composition in lambs. *Aust J Ag Res* 1991; 42: 253-267.

Capaldi ED, Sheffer J, Owens J. Food deprivation and conditioned flavor preferences based on sweetened and unsweetened foods. *Anim Learning Behav* 1991; 19: 361-368.

Collier MJ, Johnson, DF. The time window of feeding. *Physiol Behav* 1990; 48: 771-777.

Denton DA. *The Hunger for Salt –An Anthropological, Physiological and Medical Analysis*. Berlin: Springer-Verlag. 1982.

Emmans GC, Kyriazakis I. The idea of optimisation in animals: uses and dangers. *Livestock Prod Sci* 1995; 44: 189-197.

Emmans GC, Kyriazakis I. Issues arising from genetic selection for growth and body composition characteristics in poultry and pigs. In: *The Challenge of Genetic Change in Animal Production* (W.G. Hill, Bishop, S.C., McGuirk, B., McKay, J.C., Simm, G. and Webb, A.J. eds.). Occasional Publication of the British Society of Animal Science No 27, Edinburgh, pp. 39-53. 2000.

Emmans GC Kyriazakis I. Consequences of genetic change in farm animals on food intake and feeding behaviour. *Proc Nutr Soc* 2001; 60: 115-125.

Foley R. *Another Unique Species*. New York: John Wiley and Sons Inc. 1987.

Hocking PM, McCormack HA. Differential sensitivity of ovarian follicles to gonadotrophin stimulation in broiler and layer lines of domestic fowl. *J Reprod Fert* 1995; 105: 49-55.

Hughes BO. Appetites for specific nutrients. In: *Food intake regulation in poultry* (K.N. Boorman and Freeman, B.M., eds.). Proceedings of the fourteenth Poultry Science Symposium, British Poultry Science Ltd, Edinburgh, pp. 141-169. 1979.

Hughes BO, Dewar WA. A specific appetite for zinc in zinc-depleted domestic fowls. *Brit Poultry Sci* 1971; 12: 255-258.

Hutchings MR, Kyriazakis I, Gordon IJ, Jackson F. Trade-offs between nutrient intake and faecal avoidance in herbivore foraging decisions: the effect of animal parasitic status, level of feeding motivation and sward nitrogen content. *J Anim Ecol* 1999; 68: 310-323.

Kyriazakis I. The nutritional choice of farm animals: to eat or what to eat? In: *Animal Choices*, pp. 55-65. (JM Forbes, Lawrence TLJ, Rodway RG, Varley MA, eds.). Edinburgh: British Society of Animal Science. 1997.

Kyriazakis I. Emmans GC. Diet selection in pigs: choices made by growing pigs following a period of underfeeding with protein. *Anim Prod* 1991; 52: 337-346.

Kyriazakis I, Emmans GC, Whittemore CT. The ability of pigs to control their protein intake when fed in three different ways. *Physiol Behav* 1991; 50: 1197-1203.

Kyriazakis I, Leus K, Emmans GC, Haley CS, Oldham JD. The effect of breed (Large White X Landrace vs purebred Meishan) on the diets selected by pigs given a choice between two foods that differ in their crude protein contents. *Anim Prod* 1993; 56: 121-128.

Kyriazakis I, Tolkamp BJ, Emmans GC. Diet selection and animal state: an integrative framework. *Proc Nutr Soc* 1999; 58: 765-772.

Kyriazakis I, Tolkamp BJ, Hutchings MR. Towards a functional hypothesis for the occurrence of anorexia during parasitic infections. *Animal Behaviour* 1998; 56: 265-274.

Leathwood PD, Ashley DVM. Behavioural strategies in the regulation of food choice. *Experientia* Suppl. 1983; 44: 171-196.

Mela DJ, Rogers PJ. *Food, Eating and Obesity*. London: Chapman and Hall. 1998.

Moss R. Diet selection – an ecological perspective. *Proc Nutr Soc* 1991; 50: 71-75.

Murray MJ, Murray AB. Anorexia of infection as a mechanism of host defense. *Am J Clin Nutr*1979; 32: 593-596.

Nishizawa T. Some factors related to obesity in Japanese Sumo wrestlers. *Am J Clin Nutr* 1976; 29: 1168- 1174.

Provenza FD. Post-ingestive feedback as an elementary determinant of food preference and intake in ruminants. *J Range Management* 1995; 48: 2-17.

Rozin P. Sociocultural Influences on Human Food Selection. In: *Why we eat what we eat* (ED Capaldi, ed.). Washington DC: American Psychological Association, pp. 233-263. 1996.

Siegel PB. Behaviour-genetic analyses and poultry husbandry. *Poultry Sci* 1993; 72: 1-6.

Taggart N. Diet, activity and bodyweight. A study of variations in a woman. *Brit J Nutr* 1962; 16: 223-235.

Thompson SN. Physiological alterations during parasitism and their effects on host behaviour. In: *Parasitism and Host Behaviour* (CJ Barnard, Behnke JM, eds.). London, Taylor and Francis, pp. 64-94. 1990.

Tolkamp BJ, Kyriazakis I. To split behaviour into bouts, log-transform the intervals. *Anim Behav* 1999; 57: 807-817.

Trayhurn P. New insights into the development of obesity: obese genes and the leptin system. *Proc Nutr Soc* 1996; 55: 783-791.

Tremblay A, Doucet E. Influence of intense physical activity on energy balance and body fatness. *Proc Nutr Soc* 1999; 58: 99-105.

van Dam JTP. The interaction between nutrition and metabolism in West African dwarf goats infected with *trypanosomes*. PhD thesis, Wageningen Agricultural University. 1996.

Whittemore CT. *The Science and Practice of Pig Production*. Onsley Mead, Oxford: Blackwell Science Ltd. 1998.

Chapter 6

EXERCISE AND DIET-INDUCED OBESITY IN MICE

Roma R. Bell
School of Public Health, Curtin University of Technology

Abstract The mouse is a valuable experimental model for the study of diet-induced obesity in self-regulating organisms. Mice will eat a wide variety of diets and adults become obese when fed diets providing more than 30% of energy as fat. The type of fat in the diet is less important than the amount of fat. Young, growing mice do not readily become obese and obesity develops gradually in adults. Strains of mice vary in their response to dietary fat and this may result from differences in leptin and uncoupling protein response to dietary composition. Mice have high levels of spontaneous activity when given free access to exercise wheels and exhibit increased food intake, increased energy expenditure, improved aerobic capacity and enhanced ability to oxidise fat. Exercise prevents diet-induced obesity in mice fed high-fat diets and reduces body fat in mice with previously established obesity. Mice remain lean when fed high-carbohydrate diets but can only remain lean when fed high-fat diets if food intake is restricted or if they have high physical activity.

The World Health Organization reports that obesity and overweight in the human population has become a major public health concern in both developed and developing countries (WHO, 1998). While specific or multiple genetic abnormalities can cause obesity, the vast majority of overweight or obese individuals do not suffer from genetic abnormalities. This is evident from the fact that obesity rates are rising in populations whose gene pools remain relatively static. However, genetic background does influence susceptibility to weight change when exposed to various environmental conditions (WHO, 1998). The environmental conditions that

J.B. Owen et al. (eds.), Animal Models – Disorders of Eating Behaviour and Body Composition, 97–116.

have most influence on body fat content include the amount of dietary fat and the level of physical activity.

DIET AND EXERCISE METHODS FOR MICE

Laboratory animals without genetic abnormalities predisposing them to obesity are useful models to study how self-regulating organisms respond to environmental changes in diet and exercise. The mouse is a particularly useful model because its small size means that it has small energy reserves compared to its normal high energy turnover, therefore, changes in energy balance can be detected rapidly. Its small size also allows large numbers of mice to be easily housed and handled. Furthermore, the mouse readily eats diets of widely differing composition and, unlike humans, food intake is less affected by non-physiological factors.

Diets Used in Obesity Studies

Normal mice fed commercial cereal-based diets are resistant to obesity unless conditions are manipulated to alter energy balance. One approach is to induce hyperphagia by supplementing standard cereal-based diets with a variety of palatable human foods. This adds variety to the diet that increases food intake as well as increasing the fat content of the overall diet. This method of feeding is referred to as cafeteria feeding and was first used to induce obesity in rats (Sclafani & Springer, 1976). Cafeteria diets also cause substantial weight gain in mice (Bell et al., 1986; 1991). The use of cafeteria diets, as an experimental tool in obesity research, has been reviewed (Moore, 1987; Kanarek & Orthen-Gambill, 1988; Rothwell & Stock, 1988). A limitation of cafeteria feeding is the inability to control the nutrient content of the diet (Moore, 1987). Dietary intakes of protein, vitamins and minerals can vary greatly between control mice fed the cereal-based diet and the cafeteria-fed mice. Nutrient intake also varies between individual cafeteria-fed mice depending on the choice and amount of the cafeteria foods each consumes.

It is more difficult to accurately measure food intake when a variety of foods are offered since it may be difficult to separate uneaten individual foods for weighing (Kanarek & Orthen-Gambill, 1988). In addition uneaten moist foods may dry out during the time offered, resulting in an overestimation of consumption. Since one of the main features of cafeteria feeding is an increased intake of fat, it is often preferable to conduct studies on energy balance and diet-induced obesity using high-fat diets based on purified diets specifically designed for use with rodents (Reeves et al., 1993). When formulating such diets, fats should be substituted for carbohydrates on an isoenergetic basis so that all diets provide the same amount of protein, vitamins and minerals per kJ of energy. Regardless of the dietary factor being tested, the experimental group should always be compared to controls fed a similar purified diet rather than to animals fed commercial cereal-based diets.

Body Composition Analysis

Body mass and body mass changes are not reliable indicators of total body energy content or of changes in energy stores. Small changes in body fat will cause only small changes in body mass but can result in large changes in total body energy stores. Therefore it is usually desirable to assess body composition using one or more methods depending on the kind of detail required.

Chemical Analysis

Mice are small enough that it is possible to chemically analyse the whole carcass. Carcasses should be cleaned of stomach, caecum and intestinal contents, weighed and then dried to a constant weight in a convection oven or freeze-dried. Using a basic two compartment model of body composition where carcass weight = carcass fat + fat free weight, the body fat content can be determined assuming fat free mass is 73.2% water. If further detail is required the dried carcass is then ground and samples analysed for lipid, protein and total ash. Lipid content is determined gravimetrically after extraction with petroleum ether. Protein is measured by the Kjeldahl method and total ash is measured after combustion in a muffle furnace. An alternative method to drying the whole carcass involves softening it in an autoclave and then homogenising it in a small amount of water before samples are analysed for protein and lipid.

In some cases it may not be necessary to measure total body fat. Relative body fatness may be assessed easily by weighing one of the major abdominal fat pads. In female mice the weight of the retroperitoneal fat pad is positively correlated with total body fat (Bell et al., 1995). In males the weight of the epididymal fat pad is correlated with total body fat (West et al., 1992). Experimental animals must be killed in order to measure body composition by chemical analysis and this limits its use in studies where it is desirable to monitor changes in body composition over time.

Body Composition Analysis using TOBEC

Total body electrical conductivity (TOBEC) is a non-invasive method for rapidly measuring body composition. The SA-2 EM-SCAN is a small animal body composition analyser that measures the change to an electromagnetic field when an animal is placed within that field. The change is proportional to the water content of the subject and thus to its lean body mass. The animal's conductivity index is used in a prediction equation to estimate body composition. Although TOBEC technology may have some limitations with small animals (Bell et al., 1994), most research on rats has demonstrated that TOBEC predicted body fat has a good correlation with body fat determined by direct methods (Morbach & Brans, 1992; Stenger & Bielajew, 1995;) especially when prediction equations are adjusted for the population being studied (Baer et al., 1993; Bellinger & Williams, 1993). Although the body weight of the mouse is at the lower range of measurement for the SA-2 EM-SCAN instrument, it has been used successfully to measure body composition of mice using a new prediction equation developed for mice (Bell et al., 1995). Holder et al. (1999) also developed a new prediction equation that was used to successfully measure the body composition of 6500 male and female mice >10 g in mass. TOBEC measurements are rapid and non-invasive; therefore, measurements can be repeated on the same mouse over the course of the experiment.

Total body energy content can be calculated from the body fat and fat free mass or protein using the values of 39.16 kJ/g for fat and 4.99 kJ/g for lean body mass (Graham et al., 1990) or 23.6 kJ/g for protein (Tulp et al., 1979).

Exercise protocols

Various methods have been used to provide controlled amounts of exercise in experiments with mice, including swimming (Bell et al., 1984), motorised treadmills (Bulbulian et al., 1985) and motorised exercise wheels (Bell et al., 1980). While the duration and the intensity of activity can be controlled and altered to suit the experimental design, these methods may be limited by the fact that they involve forced exercise. Forced exercise itself or unintentional trauma resulting during exercise is stressful and could influence experimental outcomes. Swimming may cause cold stress (Bell et al., 1984; Harri & Kuusela, 1986) and deaths from myocardial damage and drowning have been reported (Binik & Sullivan, 1983). Mice have a very high resting metabolic rate and relatively high cost of normal activity (Koteja et al.,

1999b) so short-term forced exercise sessions may have little impact on total energy expenditure.

Providing mice with free access to running wheels is an effective way to increase voluntary activity and may often be the preferable way to exercise mice. Mice spend a great deal of time in activity wheels and voluntarily run long distances. Mice vary greatly in their use of the exercise wheels but will run from 1 km to 20 km a day with most mice averaging more than 5 km per day (Koteja et al., 1999a). Female mice (Koteja et al., 1999b; Swallow et al., 1999), like female rats (Tokuyama et al., 1982), run substantially more than males of the same species. When given access to activity wheels, mice tend to increase wheel running for a time, use then plateaus and may gradually decrease. Despite these trends in voluntary wheel running, total activity is always high compared to mice without access to running wheels and activity is sufficient to elicit physiological responses to exercise training. For example, voluntary wheel running mice have significantly higher aerobic capacity than sedentary controls (Swallow et al., 1998).

DIET-INDUCED OBESITY

Extensive research has established that diet-induced obesity is readily achieved in both male and female mice by feeding diets high in fat. The extent of the obesity depends on the level of the fat consumed and duration of the high-fat feeding and on the strain, gender and age of the mouse. Many of the obesity studies using the mouse have been done with one of several genetically obese strains. While valuable information has been gained about the mechanisms that control energy balance using these models, care should be taken when extrapolating these results to normal whole organisms that are self-regulating energy stores.

Level of Dietary Fat

Fat intake is the single most important dietary factor affecting body fat stores and the dose response relationship has been investigated in mice. Salmon and Flatt (1985) fed adult female mice diets providing 1 to 64% of energy as fat and found that body fat increased with increasing dietary fat. Although there was considerable individual variation in fat accumulation, the percentage of mice classified as obese rose sharply in mice consuming diets providing 40% or more of energy as fat. Mice fed diets containing graded amounts of lard with 6-56% energy as fat had incremental increases in body mass and fat pad mass with increasing dietary fat. Mice receiving 43% or more of energy as fat had significantly more fat compared to controls (Bourgeois et al., 1983). Female C57BL/6J mice with free access to diets providing 10, 20, 30, 40, 50 or 60% of energy from fat demonstrated increases in body weight and adipose tissue mass with each increment of dietary fat. Body fat mass was significantly greater in mice consuming more than 30% of energy as fat compared to mice receiving 10% of energy from fat (Takahashi et al., 1999). These studies indicate that diet-induced obesity is usually present in mice fed >30% of energy as fat but is seldom seen in mice fed <20% of energy as fat.

Not all strains of mice are equally prone to obesity when consuming high-fat diets (Fenton & Carr, 1951; West et al., 1992; 1995). However, even resistant strains show increases in adiposity as dietary fat is increased above 30% of energy intake (Flatt, 1991b; West et al., 1995). When thirteen strains of mice were allowed to regulate their macronutrient intake, they selected diets that provided from 26 to 83% of energy as fat and their epididymal fat mass correlated with their dietary fat intake (Smith et al., 2000). The majority of studies demonstrate that body fat stores in mice reflect the level of fat in the diet (West & York, 1998).

After having become obese by consuming high-fat diets, mice reduce body fat stores when given free access to low-fat diets rather than high-fat diets (Salmon & Flatt, 1985; Bell et al., 1995; Flatt et al., 1995b; Xin et al., 2000). These observations are further evidence that the level of dietary fat influences body fat stores.

High-fat Diets and Hyperphagia

It is generally thought that high-fat diets cause hyperphagia and that increased energy intake is the major mechanism by which high-fat diets induce obesity. Hyperphagia is not always easy to document due to the inherent difficulty in accurately measuring food intake in mice. Their total food intake is low due to their small size and there is often considerable food wastage and food contamination, making it difficult to accurately quantify food intake. Despite these limitations many studies report that high-fat diets cause hyperphagia in normal mice (Lin et al., 1979; Mercer & Trayhurn, 1987; Park et al., 1997a; Richard et al., 1988; Surwit et al., 1995; Black et al., 1998), although inbred strains show considerable variation in their response (West et al., 1992). Hyperphagia is not readily detected in young rapidly growing mice consuming high-fat diets. However, hyperphagia develops later as the animals grow and accumulate excess body fat (Lin et al., 2000). In adult mice the hyperphagia after introduction of high-fat diets may be transient with food intake initially increasing and then decreasing with time (Flatt, 1991b). Hyperphagia may develop again as fat accumulation continues (Lin et al., 2000).

High-fat diets are said to be more palatable than low-fat diets and most mice show a preference for fat (Smith et al., 2000). While food preference influences food intake in humans consuming mixed diets, it is less clear what role diet preference plays in hyperphagia in mice given a single diet of uniform composition. Cafeteria feeding increases food variety and food intake whether the cafeteria foods are high in fat or carbohydrate (Harris, 1993). Hyperphagia can result from environmental factors other than diet composition. Mice housed in groups with four per cage ate more and had greater body fat mass at all levels of dietary fat than mice housed individually while consuming the same diets (Flatt, 1995a). When mice are cafeteria-fed in a group situation there is considerable competition for favoured foods such as sunflower seeds (personal observation). Social factors may influence food intake or lead to perceived competition for food causing overconsumption when the food source is not limiting. It is likely that postingestive metabolic effects of the diet act along with diet palatability to influence food intake in animals consuming high-fat diets. Sclafani (1993) has reviewed this area.

High-Fat Diets and Energy Efficiency

Hyperphagia alone does not account for all of the fat gain in animals fed high-fat diets. There is evidence that food efficiency also increases in mice fed high-fat diets (West et al., 1992; Surwit et al., 1995). Mice fed high-fat diets have more body fat than mice fed low-fat diets even when there is no difference in food consumption (Geiger & Canolty, 1978; Smith et al., 1991; Bell et al., 1995; 1997). Adult mice fed a low-fat diet and mice pair-fed a high-fat diet had similar body weight, but the high-fat fed mice had increased body fat (Richard et al., 1988). Similar results have been reported in rats (Oscai et al., 1984; Boozer et al., 1995; Yaqoob et al., 1995). It is probable that high-fat diets influence energy balance by changes in both food intake and food efficiency.

Type of Dietary Fat

The weight of evidence supports the concept that the level of dietary fat influences the amount of body fat stores in mice. The evidence is less clear, however, on whether the composition of dietary fat exerts an independent effect on body mass and fat stores. Most of the studies have been done using rats with only a few conducted on mice. Studies generally have compared fats differing in chain length or degree of saturation and have yielded conflicting results. For example, fish oil, which is rich in long chain n-3 polyunsaturated fatty acids, has been reported to cause increased body mass (Pan & Berdanier, 1991; Yaqoob et al., 1995), decreased body fat (Su & Jones, 1993; Hill et al., 1993; Cha & Jones, 1998; Xin et al., 2000) or no change to body mass (Jones et al., 1995) compared to other dietary fats. Numerous studies have been done comparing beef tallow or lard, common food fats high in saturated fatty acids, to commonly consumed vegetable oils that are either polyunsaturated such as corn oil and safflower oil or monounsaturated such as olive oil and canola oil. Animals fed saturated fats were reported to have greater body mass and/or larger fat stores (Bourgeois et al., 1983; Mercer & Trayhurn, 1987; Takeuchi et al., 1995), to have lower body mass and/or fat stores (Pan & Storlien, 1993; Cha & Jones, 1998) or to have the same body mass and/or fat stores (Hill et al., 1993; Edwards et al., 1993; Jones et al., 1995) compared to animals fed polyunsaturated fats. Similar conflicting results were found when animals fed saturated fat were compared to animals fed monounsaturated fats. Animals fed beef fat or lard were reported to have more body fat (Takeuchi et al., 1995; Bell et al., 1997), lower body fat or body mass (Pan & Storlien, 1993) or the same body mass (Su & Jones, 1993; Jones et al., 1995) as animals fed monounsaturated fats. It is difficult to draw conclusions from these data; however, it is safe to say that under certain conditions the type of dietary fat can influence body fat stores by altering food intake or by altering mechanisms regulating fat storage or utilisation. Even when the type of fat does alter body fat mass, the effect is small compared to the effects demonstrated by changing the amount of fat in the diet.

Conjugated Linoleic Acid

Components of dietary fat may influence body fat accumulation. Conjugated linoleic acid (CLA) refers to a group of dienoic derivatives of linoleic acid formed by rumen bacteria and naturally present in some foods of animal origin. Recent evidence has demonstrated that dietary CLA reduces the body fat content of mice (Park et al., 1997b; West et al., 1998; DeLany et al., 1999; Park et al., 1999a; 1999b; DeLany & West, 2000; West et al., 2000) and prevents obesity in mice fed high-fat diets (West et al., 1998; DeLany et al., 1999). The *trans*-10,*cis*-12 isomer appears to be the form of CLA responsible for lowering body fat stores (Park et al., 1999b). Many of the studies on CLA have been done using preparations containing a mixture of

CLA isomers making it difficult to separate the biological actions of the various isomers or to establish dose response relationships. Dietary CLA decreases food intake in mice (West et al., 1998) but this is probably not its main effect on energy balance since lower concentrations of CLA that do not affect food intake still decrease body fat (DeLany et al., 1999). CLA increases energy expenditure (West et al., 2000), increases fat oxidation (Park et al., 1997b; DeLany & West, 2000), reduces the respiratory quotient (Azain et al., 2000), increases adipose tissue lipolysis and decreases lipoprotein lipase (Park et al., 1997b; 1999b). CLA has no effect on de novo fatty acid synthesis (West et al., 2000) but changes the composition of body fat. CLA decreases the expression of stearoyl-CoA desaturase leading to reduced tissue levels of monounsaturated fatty acids (Lee et al., 1998; Choi et al., 2000). CLA reduces adipose tissue cell size rather than cell number (Azain et al., 2000). These alterations in lipid metabolism contribute to the increased oxidation and decreased storage of fat in CLA-treated mice.

Dietary Protein

The level of dietary protein may influence food intake, body weight and fat stores. Mice fed low-protein diets have increased food intake (Bell et al., 1984; Toyomizu et al., 1988; West et al., 1995), increased body fat (Bell et al., 1984; West et al., 1995), increased energy expenditure (Bell et al., 1984) and decreased food efficiency (Toyomizu et al., 1988). It is possible that rodents fed diets low in protein develop hyperphagia in an attempt to consume enough protein to meet their protein requirements for lean tissue growth and maintenance. However, if protein levels are too low, both food intake and body weight decrease (Toyomizu et al., 1988; Du et al., 2000). Increasing protein intake beyond that needed for protein synthesis does not have an impact on food intake as mice eat for energy rather than protein when the protein intake is adequate (Toyomizu et al., 1988). However, high protein diets may have detrimental effects on food intake, body mass and kidney function (Geiger & Canolty, 1978: West et al., 1995). The low-protein model for the study of obesity may be less useful than the high-fat model. The protein intake range that results in hyperphagia is relatively narrow and changes with the age of the mouse (Toyomizu et al., 1988). The low-protein fed animals accumulate less fat than high-fat fed animals. Furthermore the human diets that lead to obesity are not low in protein, but rather high in fat with more than adequate protein (WHO, 1998).

Dietary Carbohydrates

Mice fed high-carbohydrate diets exhibit normal growth rates and maintain low body fat mass. There is no evidence that replacing starch in the diet with sucrose or other sugars such as glucose or fructose influences food intake or body weight gain in normal mice (Leiter et al., 1983; Seaborn & Stoecker, 1989; Surwit et al., 1995; Black et al., 1998). Studies in rats have yielded variable results when sucrose is substituted for starch in the diet (see review by Kanarek & Orthen-Gambill, 1988). However, rats given 32% sugar solutions along with their normal commercial cereal-based diet generally increase their energy intake by 15-20% and this leads to weight gain over time in adult animals. In this situation the sugar solutions are a means of inducing hyperphagia, however, the degree of hyperphagia depends on the level of protein in the diet. In rats consuming low protein diets there is less increase in energy intake than in rats consuming high levels of protein (Kanarek & Orthen-Gambill, 1988). Sugar as such does not cause obesity, but sugar can contribute to obesity if provided in conditions that promote hyperphagia.

Body Composition and Nutrient Oxidation Rates

Mice fed high-fat diets develop obesity with fat stores expanding in relation to dietary fat intake. The reason for this relationship has been explored by Flatt (1991b; 1993; 1995a; 1995b) who suggests that energy balance is really the sum of carbohydrate and fat balance. Oxidation rates of carbohydrate and fat can be determined by measuring the respiratory quotient (RQ), which is the ratio of CO_2 produced to O_2 consumed. The respiratory quotient for carbohydrate is 1.0 and that for lipid is 0.7. The RQ reflects the proportion of carbohydrate and fat oxidised. Carbohydrate oxidation depends on carbohydrate availability and oxidation increases rapidly after a meal containing carbohydrates. On the other hand, fat consumption does not stimulate fat oxidation and any unoxidised fat is stored. Fat oxidation rates are influenced by the amount of fatty acids released from fat storage and that in turn depends on the size of fat reserves and is under hormonal control. In sedentary mice fed high-fat diets, fat stores must expand until fatty acid release and oxidation rates equal fat intake. This explains the observed positive relationship between fat intake and fat stores. Recent research on leptin has suggested mechanisms that could account for this relationship.

Leptin and Diet-Induced Obesity

Leptin is a hormone produced primarily by adipose tissue and its secretion and plasma concentration are correlated with total fat mass. Leptin acts to regulate lipid reserves by decreasing food intake, increasing energy expenditure and altering fuel selection. Leptin decreases food consumption by modulating the production of hypothalamic neuropeptides. Leptin increases energy expenditure either directly or indirectly by increasing the production of triiodothyronine. Leptin alters fuel use by increasing lipid oxidation and decreasing carbohydrate oxidation. Leptin decreases body lipid stores while sparing lean body mass (Halaas et al., 1995; Kaibara et al., 1998). The physiological effects of leptin including its role in fat metabolism, the regulation of body fat mass and energy conservation during fasting have been recently reviewed (Friedman & Halaas, 1998; Ahima & Flier, 2000; Baile et al., 2000; Reidy & Weber, 2000).

Mice fed high-fat diets have increased food intake and greater body fat despite higher plasma concentrations of leptin compared to mice fed low-fat diets. These changes in body composition and leptin levels are reversible when the high-fat fed mice are changed to low-fat diets (Xin et al., 2000). C57BL/6J mice remain lean when fed low-fat diets but gain variable amounts of weight and fat when fed high-fat diets. Mice that remained lean had significantly higher plasma leptin levels and a higher ratio of plasma leptin to fat mass than the mice that became obese when fed the high-fat diet (Xin et al., 2000). The A/J mouse, like the lean C57BL/6J mouse, resists obesity even when fed high-fat diets and the A/J mouse also exhibits greater elevation of plasma leptin than can be accounted for by the increase in body fat (Surwit et al., 1997). Plasma levels of leptin are inversely correlated to subsequent changes in body weight in C57BL/6J mice challenged by a high-fat diet but not in mice fed a low-fat diet (Ahrén, 1999).

Although leptin clearly acts to regulate energy balance and body fats, normal mice challenged with high-fat diets still become obese. One reason for this may be that both peripheral and central leptin resistance develops after long-term hyperleptinaemia. Diet-induced obesity develops gradually but as fat accumulates and leptin levels increase, sensitivity to leptin administered either peripherally or centrally decreases (El-Haschimi et al., 2000; Lin et al., 2000). This suggests that even in normal mice high-fat diets can lead to leptin resistance over time. A second reason why leptin fails to prevent obesity in normal animals fed high-fat diets is suggested by the fact that dietary fat intake fails to stimulate leptin synthesis and secretion to the same extent as carbohydrate intake. Animals fed high-fat diets secrete less leptin per unit of fat mass than animals fed low-fat diets (Surwit et al., 1997; Ainslie et al.,

2000). In humans the 24-hour plasma leptin levels are 40% lower when fed high-fat meals than when fed isoenergetic low-fat meals (Havel et al., 1999). Since insulin stimulates leptin expression (see Havel, 2000), the differences in insulin secretion following fat and carbohydrate consumption may contribute to the relatively lower levels of leptin in high-fat fed animals.

The A/J mouse has low susceptibility to diet-induced obesity and a high-fat diet does not compromise leptin-dependent regulation of adipose tissue. A/J mice have higher levels of leptin and greater concentrations of uncoupling protein 1 (UCP1) in brown adipose tissue, UCP2 in white adipose tissue (Watson et al., 2000) and UCP5 in liver (Yu et al., 2000) than C57BL/6J mice that readily develop obesity when fed high-fat diets. UCPs refer to a group of related proteins expressed in various tissues that facilitate mitochondrial proton leakage and thus increase energy expenditure. Leptin is one factor that regulates UCP expression and activity (Reidy & Weber, 2000; Adams, 2000). Differences in leptin and uncoupling protein response during high-fat feeding may account for differences in strain susceptibility to diet-induced obesity.

EXERCISE AND OBESITY

The role of exercise in the regulation of food intake, body weight and body composition has been the subject of much research and controversy. Exercise contributes to the regulation of energy balance by increasing energy expenditure and regulating energy intake. Exercise alters nutrient partitioning by decreasing fat storage and enhancing fatty acid mobilisation and oxidation. Exercise can correct some of the adverse metabolic effects of obesity.

Food Intake and Energy Expenditure

Classic studies with female rats demonstrated that food intake increases as exercise duration increases over a wide range of activity levels. However, rats participating in very light exercise actually had lower food intake than sedentary rats (Mayer et al., 1954). The decrease in food intake with low levels of physical activity has not been consistently observed in other species including mice. Although Lane et al. (1991) reported decreased food intake in female mice exercising 60 minutes a day, Bulbulian et al. (1985) found no change in food intake between exercising and sedentary mice until exercise levels reach 120 or more minutes per day when exercising mice exhibited a significant increase in food intake. It is possible that exercise alters fuel partitioning by mobilising fuel stores in which case, food intake would not be changed markedly until fuel stores are reduced (Friedman, 1998).

Acute bouts of exercise do have an anorectic effect in rodents (Rivest & Richard, 1990) and such exercise is associated with short-term increases in plasma leptin concentrations but reduced adipose tissue levels of leptin mRNA (Bramlett et al., 1999). It is probable that short periods of post-exercise hypophagia have little impact on long-term energy expenditure as later food intake can readily be increased in compensation. It is difficult to quantify small changes in food intake or energy expenditure in mice. Mice have very high metabolic rates so small increases in physical activity may not increase total energy expenditure measurably. Alternately mice may reduce other forms of activity to compensate for the energy expended during short periods of exercise. However, as activity increases further there is a measurable increase in energy consumption and in energy expenditure (Yashiro & Kimura, 1979; Bulbulian et al., 1985; Bell et al., 1986; 1995; 1997). In mice with free access to exercise wheels, food consumption correlates with the number of wheel revolutions the mouse achieves (Flatt, 1991a; 1991b; Koteja et al., 1999b). Furthermore, food consumption is significantly higher when mice have free access to exercise wheels than when the same mice were denied access to the exercise wheels (Flatt, 1995a; Koteja et al., 1999b)

Body Weight and Composition

Despite the fact that high levels of physical activity lead to increased food intake, exercising mice fed low-fat diets exhibit either no change (Bell et al., 1986; Bell & McGill, 1991; Lane et al., 1991; Koteja et al., 1999b) or only small decreases in body mass or body fat (Yashiro & Kimura, 1979; Bell et al., 1980; 1995; 1997; Bulbulian et al., 1985; Swallow et al., 1998; 1999). Normal mice fed low-fat, high-carbohydrate diets maintain low body mass and low body fat even when sedentary and it would not be expected that exercise would further reduce these already low levels of body fat. However, when mice are challenged with a high-fat diet, exercise substantially decreases weight gain and prevents diet-induced obesity (Bell et al., 1991; 1995; 1997; Flatt, 1995a; Lane et al., 1991). Furthermore, exercise leads to substantial losses of fat in mice with previously established diet-induced obesity (Bell et al., 1986; 1995; Flatt, 1995a). Similar beneficial effects of exercise on body composition have been reported for rats (Tokuyama et al., 1982; Podolin et al., 1999; Melton et al., 2000). Exercising mice are able to regulate food intake to maintain low body fat regardless of their diet composition.

Exercise sufficient to decrease body fat has little effect on lean body mass in adult mice (Bulbulian et al., 1985; Bell et al., 1984; 1986; 1995; 1997; Lane et al., 1991). However, chronic voluntary wheel running increases muscle mass in 15 month-old mice (Willis et al., 1998) and may prevent age-related loss of lean body mass.

Exercise and Metabolism

Aerobic fitness is improved in mice with free access to running wheels. Exercised mice have increased VO_{2max} (MacNeil & Hoffman-Goetz, 1993; Swallow et al., 1998) and increased muscle aerobic capacity as evidenced by increased activity of cytochrome-c oxidase, citrate synthetase and carnitine palmitoyltransferase (Houle-Leroy et al., 2000). Exercising mice and rats also have an increased ability to mobilise (Podolin et al., 1999) and oxidise lipids. Exercise stimulates fat oxidation more than carbohydrate oxidation (Flatt, 1995a). Exercising mice have lower respiratory quotients than sedentary mice both during exercise (Swallow et al., 1998) and averaged over 24 hours (Flatt, 1991b). Exercise training decreases adipose tissue lipoprotein lipase and therefore decreases fat deposition. This enhanced use and decreased storage of lipids protects against excess fat accumulation when mice are fed high-fat diets.

As body fat accumulates in mice fed high-fat diets, plasma insulin is elevated and insulin resistance develops (West et al., 1995; Black et al., 1998; Ahrén, 1999; Watson et al., 2000). Since insulin influences leptin expression and both leptin and insulin are long-term regulators of food intake and energy balance (see Havel, 2000), treatments that improve insulin sensitivity would help to correct energy regulation. Exercise training prevents fat accumulation and the resulting hyperinsulinaemia (Podolin et al., 1999) and reduces insulin resistance in rats fed high-fat diets (Kraegen et al., 1989). These benefits of exercise on body weight and metabolism last only as long as the exercise program is continued.

REFERENCES

Adams SH. Uncoupling protein homologs: emerging views of physiological function. *J Nutr* 2000; 130:711-714.

Ahima RS, Flier JS. Leptin. *Annu Rev Physiol* 2000; 62:413-437.

Ahrén B. Plasma leptin and insulin in C57BL/6J mice on a high-fat diet: relation to subsequent changes in body weight. *Acta Physiol Scand* 1999; 165:233-240.

Ainslie DA, Proietto J, Fam BC, Thorburn AW. Short-term, high-fat diets lower circulating leptin concentrations in rats. *Am J Clin Nutr* 2000; 71:438-442.

Azain MJ, Hausman DB, Sisk MB, Flatt WP, Jewell DE. Dietary conjugated linoleic acid reduces rat adipose tissue cell size rather than cell number. *J Nutr* 2000; 130:1548-1554.

Baer DJ, Rumpler WV, Barnes RE, Kressler LL, Howe JC, Haines TE. Measurement of body composition of live rats by electromagnetic conductance. *Physiol Behav* 1993; 53:1195-1199.

Baile A, Della-Fera MA, Martin RJ. Regulation of metabolism and body fat mass by leptin. *Annu Rev Nutr* 2000; 20:105-127.

Bell RC, Lanou AJ, Frongillo EA Jr, Levitsky DA, Campbell TC. Accuracy and reliability of total body electrical conductivity (TOBEC) for determining body composition of rats in experimental studies. *Physiol Behav* 1994; 56:767-773.

Bell RR, James C, Vahdat M, Ho N. Voluntary exercise and energy balance in mice with diet-induced obesity. *Nutr Rep Int* 1986; 34:841-850.

Bell RR, McGill TJ. Body composition and brown adipose tissue in sedentary and active mice. *Nutr Res* 1991; 11:633-642.

Bell RR, McGill TJ, Digby PW, Bennett SA. Effects of dietary protein and exercise on brown adipose tissue and energy balance in experimental animals. *J Nutr* 1984; 114:1900-1908.

Bell RR, Spencer MJ, Sherriff JL. Diet-induced obesity in mice can be treated without energy restriction using exercise and/or a low fat diet. *J Nutr* 1995; 125:2356-2363.

Bell RR, Spencer MJ, Sherriff JL. Voluntary exercise and monounsaturated canola oil reduce fat gain in mice fed diets high in fat. *J Nutr* 1997; 127:2006-2010.

Bell RR, Tzeng DY, Draper HH. Long-term effect of calcium, phosphorus and forced exercise on the bones of mature mice. *J Nutr* 1980; 110:1161-1168.

Bellinger LL, Williams FE. Validation study of a total body electrical conductive (TOBEC) instrument that measures fat-free body mass. *Physiol Behav* 1993; 53:1189-1194.

Binik YM, Sullivan MJ. Sudden swimming deaths: a psychomotor reinterpretation. *Psychophysiology* 1983; 20:670-681.

Black BL, Croom J, Eisen EJ, Petro AE, Edwards CL, Surwit RS. Differential effects of fat and sucrose on body composition in A/J and C57BL/6J mice. *Metabolism* 1998; 47:1354-1359.

Bourgeois F, Alexiu A, Lemonnier D. Dietary-induced obesity: effect of dietary fats on adipose tissue cellularity in mice. *Br J Nutr* 1983; 49:17-26.

Boozer CN, Schoenbach G, Atkinson RL. Dietary fat and adiposity: a dose-response relationship in adult male rats fed isocalorically. *Am J Physiol (Endocrinol Metab 31)* 1995; 268:E546-E550.

Bulbulian R, Grunewald KK, Haack RR. Effect of exercise duration of feed intake and body composition of Swiss albino mice. *J Appl Physiol* 1985; 58:500-505.

Bramlett SB, Zhou J, Harris RBS, Hendry SL, Witt TL, Zachwieja JJ. Does β_3-adrenoreceptor blockade attenuate acute exercise-induced reductions in leptin mRNA? *J Appl Physiol* 1999; 87:1678-1683.

Cha MC, Jones PJH. Dietary fat type and energy restriction interactively influence plasma leptin concentration in rats. *J Lipid Res* 1998; 39:1655-1660.

Choi Y, Kim Y-C, Han Y-B, Park Y, Pariza MW, Ntambi JM. The *trans*-10,*cis*-12 isomer of conjugated linoleic acid downregulates stearoyl-CoA desaturase 1 gene expression in 3T3-L1 adipocytes. *J Nutr* 2000; 130:1920-1924.

DeLany JP, Blohm F, Truett AA, Scimeca JA, West DB. Conjugated linoleic acid rapidly reduces body fat content in mice without affecting energy intake. *Am J Physiol (Regulatory Integrative Comp Physiol 45)* 1999; 276: R1172-R1179.

DeLany JP, West DB. Changes in body composition with conjugated linoleic acid. *J Am Coll Nutr* 2000; 19:487S-493S.

Du F Higginbotham DA, White BD. Food intake, energy balance and serum leptin concentrations in rats fed low-protein diets. *J Nutr* 2000; 130:514-521.

Edwards MS, Smith BA, Kainer RA, Sutherland TM. Effect of dietary fat and aging on adipose tissue cellularity in mice differing in genetic predisposition to obesity. *Growth Dev Aging* 1993; 57:45-51.

El-Haschimi K, Pierroz DD, Hileman SM, Bjørbæk C, Flier JS. Two defects contribute to hypothalamic leptin resistance in mice with diet-induced obesity. *J Clin Invest* 2000; 105:1827-1832.

Fenton PF, Carr CJ. The nutrition of the mouse XI. Response of four strains to diets differing in fat content. *J Nutr* 1951; 45:225-233.

Flatt JP. Opposite effects of variations in food intake on carbohydrate and fat oxidation in ad libitum fed mice. *J Nutr Biochem* 1991a; 2:186-192.

Flatt JP. The RQ/FQ concept and body weight maintenance. *Nestle Foundation Ann Rep* 1991b; 49-71.

Flatt JP. Dietary fat, carbohydrate balance, and weight maintenance. *Ann NY Acad Sci* 1993; 683:122-140.

Flatt JP. McCollumn Award Lecture: Diet, lifestyle, and weight maintenance. *Am J Clin Nutr* 1995a; 62:820-836.

Flatt JP. Body composition, respiratory quotient, and weight maintenance. Am J Clin Nutr 1995b; 62:1107S-1117S.

Friedman MI. Fuel partitioning and food intake. *Am J Clin Nutr* 1998; 67(suppl):513S-518S.

Friedman JM, Halaas JL. Leptin and the regulation of body weight in mammals. *Nature* 1998; 395:763-770.

Geiger LW, Canolty NL. Influence of dietary protein concentration upon energy utilization in mice fed diets containing varying levels of fat and carbohydrate. *J Nutr* 1978; 108:1540-1545.

Graham R, Chang S, Lin D, Yakubu F, Hill JO. Effect of weight cycling on susceptibility to dietary obesity. *Am J Physiol (Regulatory Integrative Comp Physiol 28)* 1990; 259:R1096-R1102.

Halaas JL, Gajiwala KS, Maffei M, Cohen SL, Chait BT, Rabinowitz D, Lallone RL, Burley SK, Friedman JM. Weight-reducing effects of the plasma protein encoded by the *obese* gene. *Science* 1995; 269:543-546.

Harri M, Kuusela P. Is swimming exercise or cold exposure for rats. *Acta Physiol Scand* 1986; 126:189-197.

Harris RBS. The impact of high- or low-fat cafeteria foods on nutrient intake and growth of rats consuming a diet containing 30% energy as fat. *Int J Obes* 1993; 17:307-315.

Havel PJ. Role of adipose tissue in body-weight regulation: mechanisms regulating leptin production and energy balance. *Proc Nutr Soc* 2000; 59:359-371.

Havel PJ, Townsend R, Chaump L, Teff K. High fat meals reduce 24 h circulating leptin concentrations in women. *Diabetes* 1999; 48:334-341.

Hill JO, Peters JC, Lin D, Yakubu F, Greene H, Swift L. Lipid accumulation and body fat distribution is influenced by type of dietary fat fed to rats. *Int J Obes* 1993; 17:223-236.

Holder RB, Moura AS, Lamberson WR. Direct and correlated responses to selection for efficiency of lean gain in mice. *J Anim Sci* 1999; 77:575-581.

Houle-Leroy P, Garland T Jr, Swallow JG, Guderley H. Effects of voluntary activity and genetic selecion on muscle metabolic capacities in huse mice *Mus domesticus*. *J Appl Physiol* 2000; 89:1608-1616.

Jones PHJ, Toy BR, Cha MC. Differential fatty acid accretion in heart, liver and adipose tissues of rats fed beef tallow, fish oil, olive oil and safflower oils at three levels of energy intake. *J Nutr* 1995; 125:1175-1182.

Kaibara A, Moshyedi A, Auffenberg T, Abouhamze A, Copeland EM III, Kalra S, Moldawer LL. Leptin produces anorexia and weight loss without inducing an acute phase response or protein wasting. *Am J Physiol (Regulatory Integrative Comp Physiol 43*) 1998; 274:R1518-R1525.

Kanarek Robin B, Orthen-Gambill Nilla. "Dietary-induced obesity in experimental animals." In *Use of Animal Models for Research in Human Nutrition*, AC Beyen, CE West, eds. Basel: Karger, 1988.

Koteja P, Garland T Jr, Sax JK, Swallow JG, Carter PA. Behaviour of house mice artivicially selected for high levels of voluntary wheel running. *Anim Behav* 1999a; 58:1307-1318.

Koteja P, Swallow JG, Carter PA, Garland T Jr. Energy cost of wheel running in house mice: implications for coadaptation of locomotion and energy budgets. *Physiol Biochem Zoo* 1999b; 72:238-249.

Kraegen EW, Storlien LH, Jenkins AB, James DE. Chronic exercise compensates for insulin resistance induced by a high-fat diet in rats. *Am J Physiol (Endocrinol Metab 19)* 1989; 256:E242-E249.

Lane HW, Keith RE, Strahan S, White MT. The effect of diet, exercise and 7,12-dimethylbenz(a)anthracene on food intake, body composition and carcass energy levels in virgin female BALB/c mice. *J Nutr* 1991; 121:1876-1882.

Lee KN, Pariza MW, Ntambi JM. Conjugated linoleic acid decreases hepatic stearoyl-CoA desaturase mRNA expression. *Biochem Biophys Res Commun* 1998; 248:817-821.

Leiter EH, Coleman DL, Ingram DK, Reynolds MA. Influence of dietary carbohydrate on the induction of diabetes in C57BL/KsJ-db/db diabetes mice. *J Nutr* 1983; 113:184-195.

Lin P-Y, Romsos DR, Vander Tuig JG, Leveille GA. Maintenance energy requirements, energy retention and heat production of young obese (*ob/ob*) and lean mice fed a high-fat or a high-carbohydrate diet. *J Nutr* 1979; 109:1143-1153.

Lin S, Thomas TC, Storlien LH, Huang XF. Development of high fat diet-induced obesity and leptin resistance in C57Bl/6J mice. *Int J Obes Relt Metab Disord* 2000; 24:639-646.

MacNeil B, Hoffman-Goetz L. Exercise training and tumor metastasis in mice: influence of time and exercise onset. *Anticancer Res* 1993; 13:2085-2088.

Mayer J, Marshall NB, Vitale JJ, Cristensen JH, Mashayekhi MB, Stare FJ. Exercise, food intake and body weight in normal rats and genetically obese adult mice. *Am J Physiol* 1954; 177:544-548.

Melton SA, Hegsted M, Keenan MJ, Zhang Y, Morris S, Potter Bulot L, O'Neil CE, Morris GS. Swimming eliminates the weight gain and abdominal fat associated with ovariectomy in the retired breeder rat despite high-fat diet selection. *Appetite* 2000; 35:1-7.

Mercer SW. Trayhurn P. Effect of high fat diets on energy balance and thermogenesis in brown adipose tissue of lean and genetically obese ob/ob mice. *J Nutr* 1987; 117:2147-2153.

Moore BJ. The cafeteria diet—an inappropriate tool for studies of thermogenesis. *J Nutr* 1987; 117:227-231.

Morbach CA, Brans YW. Determination of body composition in growing rats by total body electrical conductivity. *J Pediatr Gastroenterol Nutr* 1992; 14:283-292.

Oscai LB, Bjrown MM, Miller C. Effect of dietary fat on food intake, growth and body composition in rats. *Growth* 1984; 48:415-424.

Pan DA, Storlien LH. Dietary lipid profile is a determinant of tissue phospholipid fatty acid composition and rate of weight gain in rats. *J Nutr* 1993; 123:512-519.

Pan JS, Berdanier CD. Dietary fat saturation affects hypatocyte insulin binding and glucose metabolism in BHE rats. *J Nutr* 1991; 121:1820-1826.

Park EI, Paisley EA, Mangian HJ, Swartz DA, Wu M, O'Morchoe PJ, Behr SR, Visek WJ, Kaput J. Lipid level and type alter stearoyl CoA desaturase mRNA abundance differently in mice with distinct susceptibilities to diet-influenced diseases. *J Nutr* 1997a; 127:566-573.

Park Y, Albright KJ, Liu W, Storkson JM, Cook ME, Pariza MW. Effect of conjugated linoleic acid on body composition in mice. *Lipids* 1997b; 32:853-858.

Park Y, Albright KJ, Storkson JM, Liu W, Cook ME, Pariza MW. Changes in body composition in mice during feeding and withdrawal of conjugated linoleic acid. *Lipids* 1999a; 34:243-248.

Park y, Storkson JM, Albright KJ, Liu W, Pariza MW. Evidence that the *trans*-10,*cis*-12 isomer of conjugated linoleic acid induces body composition changes in mice. *Lipids* 1999b; 34:235-241.

Podolin DA, Wei Y, Pagliassotti MJ. Effects of a high-fat diet and voluntary wheel running on gluconeogenesis and lipolysis in rats. *J Appl Physiol* 1999; 86:1374-1380.

Reeves PG, Nielsen FH, Fahey GC., Jr. AIN-93 Purified diets for laboratory rodents: Final report of the American Institute of Nutriton Ad Hoc Writing Committee on the Reformulation of the AIN-76A Rodent Diet. *J Nutr* 1993; 123:1939-1951.

Reidy SP, Weber J-M. Leptin: an essential regulator of lipid metabolism. *Comp Biochem Physiol* 2000; 125:285-297.

Richard D Boily P, Dufresne MC, Lecompte M. Energy balance and facultative diet-induced thermogenesis in mice fed a high-fat diet. *Can J Physiol Pharmacol* 1988; 66:1297-1302.

Rivest S, Richard D. Involvement of corticotropin-releasing factor in the anorexia induced by exercise. *Brain Res Bull* 1990; 25:169-172.

Rothwell NJ, Stock MJ. The cafeteria diet as a tool for studies of thermogenesis. *J Nutr* 1988; 118:925-928.

Salmon DMW and Flatt JP. Effect of dietary fat content on the incidence of obesity among ad libiturm fed mice. *Int J Obes* 1985; 9:443-449.

Seaborn CD, Stoecker BJ. Effects of starch, sucrose, fructose and glucose on chromium absorption and tissue concentrations in obese and lean mice. *J Nutr* 1989; 119:1444-1451.

Sclafani, Anthony. "Dietary Obesity." I*n Obesity Theory and Therapy* 2nd Edition. Albert J Stunkard and Thomas A. Wadden, eds. New York, NY: Raven Press. 1993.

Sclafani, A, Springer D. Dietary obesity in adult rats: similarities to hypothalamic and human obesity syndromes. *Physiol Behav* 1976; 18:1021-26.

Smith BA, Edwards MS, Ballachey BE, Cramer DA, Sutherland Tm. Body weight and longevity in genetically obese and non-obese mice fed fat-modified diets. *Growth Dev Aging* 1991; 55:81-89.

Smith BK, Andrews PK, West DB. Macronutrient diet selection in thirteen mouse strains. *Am J Physiol (Regulatory Integrative Comp Physiol)* 2000; 278:R797-R805.

Stenger J, Bielajew C. Comparison of TOBEC-derived total body fat with fat pad weights. *Physiol Behav* 1995; 57:319-323.

Su W, Jones PJH. Dietary fatty acid composition influences energy accretion in rats. *J Nutr* 1993; 123:2109-2114.

Surwit RS, Feinglos MN, Rodin J, Sutherland A, Petro AE, Opara EC, Kuhn CM, Rubuffé-Scrive M. Differential effects of fat and sucrose on the development of obesity and diabetes in C57BL/6J and A/J mice. *Metabolism* 1995; 44:645-651.

Surwit RS, Petro AE, Parekh P, Collins S. Low plasma leptin in response to dietary fat in diabetes- and obesity-prone mice. *Diabetes* 1997; 46:1516-1520.

Swallow JG, Garland T Jr, Cater PA, Zhan WZ, Sieck GC. Effects of voluntary activity and genetic selection on aerobic capacity in house mice *(Mus domesticus)*. *J Appl Physiol* 1998; 84:69-76.

Swallow JG, Koteja P, Carter PA, Garland T Jr. Artificial selection for increased wheel-running activity in house ice results in decreased body mass at maturity. *J Exp Biol* 1999; 202:2513-2520.

Takahashi M, Ikemoto S, Ezaki O. Effect of the fat/carbohydrate ratio in the diet on obesity and oral glucose tolerance in C57BL/6J mice. *J Nutr Sci Vitaminol (Tokyo)* 1999; 45:583-593.

Takeuchi H, Matsuo T, Tokuyama K, Shimomura Y, Suzuki M. Diet-induced thermogenesis is lower in rats fed a lard diet than in those fed a high oleic acid safflower oil diet, a safflower oil diet or a linseed oil diet. *J Nutr* 1995; 125:920-925.

Tokuyama K, Saito M, Okuda H. Effects of wheel running on food intake and weight gain of male and female rats. *Physiol Behav* 1982; 28:899-903.

Toyomizu M, Hayashi K, Yamashita K, Tomita Y. Response surface analyses of the effects of dietary protein on feeding and growth patterns in mice from weaning to maturity. *J Nutr* 1988; 118:86-92.

Tulp OL, Krupp PP, Danforth E Jr, Horton ES. Characteristics of thyoid function in experimental protein malnutrition. *J Nutr* 1979; 109:1321-1332.

Watson PM, Commins SP, Beiler RJ, Hatcher HC, Gettys TW. Differential regulation of leptin expression and function in A/J vs. C57BL/6J mice during diet-induced obesity. *Am J Physiol (Endrocrinol Metab)* 2000; 279:E356-E365.

West DB, Blohm FY, Truett AA, DeLany JP. Conjugated linoleic acid persistently increases total energy expenditure in AKR/J mice without increasing uncoupling protein gene expression. *J Nutr* 2000; 130:2471-2477.

West DB, Boozer CN, Moody DL, Atkinson RL. Dietary obesity in nine inbred mouse strains. *Am J Physiol (Regulatory Interative Comp Physiol 31)* 1992; 262:R1025-R1032.

West DB, DeLany JP, Camet PM, Blohm F, Truett AA, Scimeca J. Effects of conjugated linoleic acid on body fat and energy metabolism in the mouse. *Am J Physiol (Regulatory Integrative Comp Physiol 44)* 1998; 275:R667-R672.

West DB, Waguespack J, McCollister S. Dietary obesity in the mouse: interaction of strain and diet composition. *Am J Physiol (Regulatory Integrative Comp Physiol 37)* 1995; 268:R658-R665.

West DB, York B. Dietary fat, genetic predisposition, and obesity: lessons from animal models. *Am J Clin Nutr.* 1998; 67(suppl):505S-512S.

Willis PE, Chadan SG, Baracos V, Parkhouse WS. Restoration of insulin-like growth factor I action in skeleton muscle of old mice. *Am J Physiol (Endocrinol Metab 38)* 1998; 275:E525-530.

World Health Organization. *Obesity preventing and managing the global epidemic. Report of a WHO Consultation on Obesity*. Geneva: World Health Organization. 1998.

Xin X, Storlien LH, Huang XF. Hypothalamic c-*fos*-like immunoreactivity in high-fat diet-induced obese and resistant mice. *Brain Res Bull* 2000; 52:235-242.

Yu XX, Mao W, Zhong A, Schow P, Brush J, Sherwood SW, Adams SH, Pan G. Characterization of novel UCP5/BMCP1 isoforms and differential regulation of UCP4 and UCP5 expression through dietary or temperature manipulation. *FASEB* J 2000; 14:1611-1618.

Yaqoob P, Sherrington EJ, Jeffery NM, Sanderson P, Harvey DJ, Newsholme EA, Calder PC. Comparison of the effects of a range of dietary lipids upon serum and tissue lipid composition in the rats. *Int J Biochem Cell Biol* 1995; 27:297-310.

Yashiro M, Kimura S. Effect of voluntary exercise on physiological function and feeding behavior of mice on a 20% casein diet or a 10% casein diet. *J Nutr Sci Vitaminol (Tokyo)* 1979; 25:23-32.

PART 3

Genetic models of animal obesity

Chapter 7

THE OBESITY (*ob*) GENE AND LEPTIN IN ANIMAL MODELS OF OBESITY

Karen A. Augustine-Rauch
GlaxoSmithKline Pharmaceuticals

Abstract Leptin, the product of the *ob* (obesity) gene, is a hormone produced from adipocyte tissue and acts upon the central nervous system to regulate activities involved in appetite regulation and energy expenditure. Demonstration that administration of leptin to *ob* mice resulted in dramatic weight reduction led to an unprecedented $20 million purchase price for corporate patent ownership of the leptin gene. However, monogenic forms of obesity produced in various rodent models do not necessarily reflect the genetic background of obese human subjects. Clinical treatment of human obesity is challenging due to its complex etiology involving behavior, energy expenditure, and genetics. In this chapter, leptin and its involvement in the genetics and physiology of obesity will be discussed, including its therapeutic potential in the clinical treatment of this health condition.

Introduction

Obesity is a major heath issue in many Westernized countries. In the United States, more than half of American adults are classified as overweight and almost one quarter of these individuals are clinically obese. Placing second to smoking, obesity is the greatest cause of mortality in the U.S. Most of these deaths are manifested as obesity-related diseases including non-insulin-dependent diabetes mellitus (NIDDM or Type II diabetes), atherosclerosis and

J.B. Owen et al. (eds.), Animal Models – Disorders of Eating Behaviour and Body Composition, 119–131.

hypertension. Altogether, these diseases are not only a principal cause of excess mortality but also are a major cause of morbidity and decreased quality of life, particularly amongst middle age and older individuals (Echwald et al.,1999). The mechanisms behind obesity remain to be defined. Although excessive food intake and decreased energy expenditure lead to a metabolic partitioning of excess energy into fat, and this general imbalance does explain the underlying physiological basis of obesity, the factors which cause this imbalance are more complex.

With the rapidly increasing spread in obesity in Westernized societies, obesity is commonly considered a lifestyle disease. Modernization has lead to significant lifestyle changes including motorized transportation and more sedentary forms of work and recreational activities, together resulting in general reduction of physical activity. Lower physical activity combined with economical high-fat and calorie-containing processed foods likely do contribute to the significant increases in obesity. Behavioral and socio-economic factors also appear to contribute to this health problem. Obese subjects have been demonstrated to eat more food, particularly foods rich in fat, rather than just having a reduced metabolic turnover. Morbid obesity is associated with depression, where depression appears to be part of the pathogenesis of the disease rather than of the obese state. Also, there appears to be a differential response to obesity, where obese subjects gain weight during depression, while leaner subjects lose weight during this state. Poverty is one of the strongest predictors of obesity, particularly amongst women, indicating a considerable impact of environmental influences contribute to this health problem and such factors are also very difficult to define and control for in scientific studies.

Interestingly, even in obese subjects, the weight gain throughout life is very small compared to the energy expended. For instance, a typical American woman may gain 11 kg between the ages of 25 and 65, but this is the result of only 350 mg of excess daily food on the average. In consideration that more than 18 tons of food are ingested over the 40 year period, this means that a weight gain of 11 kg indicates that the regulation of the fat depots would be in error by less than .03% (Houseknecht et al., 1998). Current research strongly indicates that hereditary factors could be of equal importance for the development of both obesity and obesity-related diseases. From an evolutionary perspective, mutations impacting insulin sensitivity and energy expenditure may have been advantageous for survival during periods of famine in a hunter-gatherer or primitive agrarian society. However in today's society, these genetic mutations, often called 'thrifty genes', have instead become detrimental due to their contribution to excessive fat accumulation, obesity and obesity-related diseases (Lonnqvist et al., 1999). Studies of twins and adopted children have shown that most of familial aggregation of obesity is due to genetic influence rather than to shared family environment. Studies performed on monozygotic twins reared apart and together have indicated that the influence of heritability may be as large as 60-80% and that the twins tend to demonstrate a similar body weight development and energy expenditure rates (Stunkard et al, 1986; Vogler et al., 1995). However, in single and complex segregation analysis, heredity could only explain about 30% of the variability in body mass index (BMI) (Bouchard and Perusse, 1993).

Several genes have been identified in rodents which when mutated, cause obesity in mice and/or rats. Identification and characterization of these gene products have led to a better understanding of biochemical pathways underlying weight homeostasis and obesity. Human homologues of some of these genes have also been identified, where on rare occasions, single–gene mutations leading to massive obesity have now been described in human subjects. For many years, it has been suspected that secondary to excessive food intake, the storage of additional adipose tissue will give signals to the brain that the body is obese, which in turn, will make the subject eat less and burn more fuel. The discovery of leptin, a protein secreted from adipocytes which acts principally on its receptors in the hypothalamus, appears to be at least one protein which is involved in this feedback system. In humans and animals, leptin expression has been found to positively correlate with body fat percentages and body mass index (BMI). Additional studies have demonstrated that leptin and its receptor are components of a hormonal feedback loop regulating adipose mass through their influence on food intake and energy expenditure.

History and Discovery of Leptin and the Leptin Receptor

One of the early rodent models of obesity was discovered by animal caretakers at the Jackson laboratories in 1950 (Ingalls et al., 1950). This mouse model was called *ob/ob* (*obesity*). The *ob/ob* mice have a recessive genetic obesity that results in sterile adult mice with over 50% adipose tissue. These obese mice eat continuously and have a weight three times that of normal mice. Later, a similar mutant mouse strain, *db/db* (*diabetes*), was discovered which was also obese as well as hyperglycemic (Hummel et al., 1966).

Not long after the discovery of the *db/db* mice, Coleman undertook seminal parabiosis experiments of the *ob/ob* and *db/db* mice. In these studies, two mice were surgically joined to permit cross-circulation in that *db/db* and *ob/ob* mutants were joined to wild-type mice as well as to each other. Coleman found that there was reduced food intake and body weight in the *ob/ob* mice which shared circulation with the wild-type or *db/db* mutants, whereas there were no responses in the *db/db* mutants which shared circulation with wild-types or *ob/ob* mutants. Coleman concluded that *ob/ob* mice fail to make a circulating factor required for regulating food intake, however their brains can still respond to the factor if it became present in the circulation. He also concluded that the *db/db* mice make the circulating factor but their brains can not respond it (Coleman, 1973).

It took over 20 years to verify Coleman's hypothesis. This was accomplished by using molecular approaches to definitively identify the genetic defect in the *ob/ob* mice. A team led by Jeffrey Friedman at Rockefeller University used positional cloning to identify and sequence both the mouse *obese* gene and its human homologue (Zhang et al., 1994). The *ob* gene is expressed exclusively in white and brown adipose tissue and encodes a 167 amino acid-long, 16 killidalton (kDa) plasma protein with striking structural similarities to members of the long-chain helical cytokine family. This protein was named leptin. Under normal conditions, leptin mRNA levels correlate with the amount of body fat. However, in the homozygote *ob/ob* mice, two different mutations have been demonstrated. The first *ob* mutant essentially possesses a null mutation, resulting in failure of production of leptin mRNA. The second mutation harbors a nonsense mutation in the protein coding region, leading to the production of a truncated, inactive form of leptin which also results in a compensatory 20-fold increase in defective leptin mRNA. Together these studies demonstrate that obesity in the mutant *ob/ob* mice can be attributed to a deficiency in active leptin, which results in increased appetite, decreased energy expenditure and weight gain. The mutation in the *diabetes* mouse mutant was also characterized in the mid-1990's. A high-affinity leptin receptor was cloned from mouse choroid plexus and was mapped to the same region of mouse chromosome 4 which contains the *db* locus. A mutation in the *diabetes* locus results in abnormal alternative splicing of the leptin receptor mRNA resulting in a defective receptor protein. This receptor possesses a truncated intracellular domain and can no longer conduct cytosolic signaling in response to leptin. In the *db* mutant, this defective signaling mechanism leads to a compensatory increase in *ob* gene expression and plasma leptin levels (Gwo-Hwa et al., 1996).

Administration of leptin to wild-type and *ob* mice results in dramatic weight reduction. After 4 days of daily leptin injection, morbidly obese *ob* mice consumed 60% less food compared to untreated *ob* control mice. Following 4 weeks of leptin treatment, the obese mice had presented 40% weight reduction due to a 75% reduction in fat mass. In addition, the animals exhibited increased physical activity along with decreased serum insulin and glucose levels (Pelleymounter et al., 1995). Lean wild-type mice also respond to leptin, where the animals exhibit a 12% body weight reduction by 4 weeks of treatment (Haalas et al 1995). Of interest was the finding that female sterility in the *ob* mice was corrected by leptin treatment and that leptin administration prevents starvation-induced changes in the gonadal, adrenal and thyroid axes in mice (Ahima et al., 1996). These findings suggest that leptin may have a survival-based function which impacts female reproductive capacity, whereby a marked reduction of circulating leptin levels during starvation would prevent ovulation and pregnancy in situations where the animal has limited access to food (Lonnqvist et al., 1999).

Although loss-of-function mutations in leptin and its receptor gene have been demonstrated to be the basis of obesity in certain rodent strains, these findings have not led to an explanation of any prevalent form of human obesity. Only two human families have been found where mutations in leptin were associated with obesity: a truncation mutation in the leptin gene of two Pakistani cousins and a single Arg to Trp missense mutation in the leptin gene of three morbidly obese members of an extended Turkish family (Montague et al., 1997). In addition, five single-step mutations identified in human leptin receptors have not been found to be associated with obesity (Strobel et al., 1998).

Function of the Leptin Receptor

Expression of the leptin receptor gene undergoes alternative splicing to produce several splice variants of the receptor which are designated as OB-Ra, OB-Rb, OB-Rc, OB-Rd and OB-Re. The receptors have an extracellular domain of 840 amino acids, a transmembrane domain of 34 amino acids and a variable intracellular domain, which is characteristic for each of the five receptor isoforms (reviewed by Fruhbeck et al., 1998). Only the "long" form of the receptor (OB-Rb or Ob-RL) is expressed in various regions of the brain and is thought to be responsible for the primary actions of leptin. In contrast, OB-Re, which lacks the intracellular domain, encodes a soluble receptor which may function as a specific binding protein. OB-Re has been reported to be expressed in mice at sufficient levels to act as a buffering system for circulating leptin. Others have proposed that some of the receptor isoforms are involved in the transport of leptin in blood and in its crossing of the blood-brain barrier.

The leptin receptor is similar to the gp130 signal transduction arm of class I cytokine receptor family members which include IL-6, G-CSF and LIF. Class I cytokine receptors typically lack intrinsic tyrosine kinase activity and are activated by formation of homo- or heterodimers. Leptin receptors have been reported to form homodimers. Upon binding of the long form of the receptor by leptin, receptor aggregation occurs followed by phosphorylation of amino acid residues on the receptor intracellular domain which result in the activation of the kinases, Janus kinases (JAK). JAK are associated with membrane-proximal sequences of the receptor intracellular domain. JAK then phosphorylates specific receptor tyrosine residues that provide docking sites for members of the signal transducers and activators of transcription (STAT) family. Phosphorylation of the STAT by JAK results in dimerization of STAT proteins and their translocation to the nucleus enabling them to undertake their regulatory function as transcriptional factors (reviewed by Houseknecht et al., 1998).

It is possible that various isoforms of the leptin receptor may conduct signal transduction by pathways other than the JAK-STAT system. Although the long form of the leptin receptor is primarily expressed in the hypothalamus, other isoforms are more ubiquitously expressed and may signal through various pathways. Signal transduction by other members of the class I cytokine receptor family have been found to be also linked to the mitogen-activated protein kinase (MAPK) and phosphotidyl inositol-3 (PI-3) kinase pathways.

Leptin in Energy Balance Regulation

Food intake and energy expenditure are regulated by various hormones including insulin, leptin, satiety peptides and glucocorticoids, which are secreted from organs into the bloodstream and act upon neurons in the hypothalamus. Stimulated hypothalamic neurons signal autonomic neurons that regulate eating behavior, metabolism and energy production (reviewed by Augustine and Rossi, 1999). Anabolic and catabolic metabolic pathways maintain the size of fat mass in this process. Anabolic pathways increase food intake and energy storage, whereas catabolic pathways decrease food intake and energy storage and a delicate balance between these pathways exists in maintaining weight homeostasis.

Leptin and insulin play a role in catabolism. When food intake and energy utilization are in balance, leptin and insulin are produced into the circulation and the factors successfully stimulate receptors on hypothalamic neurons. As the adipose tissue mass increases in obese humans and animals, circulating levels of insulin and leptin also increase. Surprisingly, increases in serum leptin levels do not result in promoting catabolic processes but instead, there is a decrease in the responsiveness of the hypothalamic neurons to leptin, leading to a metabolic shift from catabolism to anabolism, thus resulting in increased appetite and lower energy expenditure. Likewise, excessive circulating insulin leads to a similar decrease in responsiveness of the insulin receptor on targeted tissues including the brain, adipose, liver and skeletal muscles. Hyperinsulinemia and insulin resistance also lead to obesity as well as Type II diabetes.

The role of insulin on leptin regulation is still in debate. Leptin may have an influence on both insulin secretion as well as resistance, whereas insulin may influence the expression and secretion of leptin in fat. In human hepatic HepG2 cells, leptin has been demonstrated to inhibit several insulin-stimulated signaling activities, such as tyrosine phosphorylation of insulin receptor substrate-1 (IRS-1) and downregulation of gluconeogenesis. Leptin administration to pancreatic islet cells isolated from *ob/ob* mice but not *db/db* mice produced a dose-dependent inhibition of glucose-stimulated insulin secretion. These data suggest that in rodents, a functional leptin receptor is present on pancreatic islets and that leptin production (particularly from abdominal adipose tissue) may modify insulin secretion (reviewed by Lonnqvist et al., 1999).

Subtle changes in energy balance have significant effects on leptin expression. As little as a 10% weight reduction in obese human subjects results in a 53% reduction in serum leptin. Moreover, a 10% increase in body weight causes a 300% increase in serum leptin (Considine et al., 1996; Kolacynski et al., 1996b). In this regard, leptin functions as an "adipostat", where in conditions of homeostatic energy balance, leptin signals to the brain the status of body energy stores, serving as an afferent signal to the brain for controlling body weight. Conversely, in conditions of energy imbalance, leptin also serves as a sensor to signal the body to reestablish balance between food intake and energy expenditure.

Leptin treatments in rodents can be utilized to eliminate all visible fat and the loss of body weight and fat deposits are not repleted for several weeks following termination of treatment. These responses suggest that leptin reduces food intake as well as enhancing metabolic rate, which is in contrast to the reduced metabolic rate associated with limited feeding. A leptin-induced increase in thermogenesis, oxygen consumption, uncoupling protein (UCP) mRNA expression and motor activity has been observed together with a normalization in serum glucose and insulin levels.

Exercise studies have demonstrated species differences in leptin response between rodents and humans. In rats, *ob* mRNA levels were reduced immediately after an acute period of exercise. However, no effect on leptin concentrations were observed following a 20-mile run by highly trained male athletes or following 60 minutes of moderate-intensity aerobic exercise in either lean or obese subjects (reviewed by Fruhbeck et al., 1998). The same lack of effect was also observed in exercise studies in post-menopausal women and in sedentary adults, suggesting that there is little evidence from studies on human subjects that increases in energy expenditure from increased activity alter leptin levels independently of the effects on adipose tissue mass.

Mechanisms of Leptin Action: CNS versus Peripheral Effects

Leptin is produced and secreted from white adipocytes into the bloodstream where it is transported to the brain. By stimulating receptors in the brain, leptin acts to stimulate or inhibit the release of factors that ultimately result in a reduction in food intake, an increase in energy expenditure and increased physical activity. Leptin and potentially additional factors, also function in a negative feedback loop where it acts upon adipocytes to inhibit further expression of the leptin (*Ob*) gene. Although additional effectors of leptin action in the brain remain to be identified, several central nervous system (CNS) factors and receptors have been established as key downstream components of leptin action including NPY, CRH, MSH and the MC-4 receptor. NPY appears to be an important mediator in the response to starvation, whereas CRH, MSH and MC-4 are involved in the hypothalamic response to obesity.

Neuropeptide Y (NPY) has been identified as a major target of leptin action. NPY is expressed in the hypothalamus and stimulates food intake and inhibits brown fat thermogenesis as well as increases plasma insulin and corticosteroid levels. Leptin treatment lowers NPY levels in *ob/ob* mice as well as in normal animals. When NPY gene expression was deleted in *ob/ob* mice, absence of NPY resulted in an attenuation of all aspects of the obesity phenotype in the *ob/ob* mice, although it did not completely rescue the phenotype. In addition, NPY knockout mice generated on a non-obese strain background exhibited normal food intake, energy expenditure and response to leptin (Houseknecht et al., 1998). Together, these results suggest that although NPY may have a central role in leptin action, other neuroendocrine targets of leptin are likely to exist.

In contrast to leptin inhibiting NPY expression, leptin stimulates expression of the corticotrophin-releasing hormone (CRH) gene. CRH plays a role in catabolism, where production of this hormone will reduce appetite and increase energy expenditure. Leptin's counter-mediated roles in regulating NPY and CRH expression may play an important role in determining the state of energy balance.

Recent studies of another mouse obesity mutant, *agouti*, have demonstrated that defective signaling of the melanocortin-4 (MC-4) receptor and its ligand, melanocyte-stimulating hormone (MSH) have a mechanistic role in the pathogenesis of the *agouti* obesity syndrome. Both *agouti* and MC-4 knockout mutants exhibit high plasma leptin levels. Additional studies have demonstrated that MC-4 receptor signaling is necessary to mediate leptin's effect on food intake and body weight, indicating that the obesity phenotype, resulting from disruption in MC-4 signaling, is caused by the loss of an important downstream target in the leptin signaling cascade (Seeley et al, 1997).

Leptin may act on peripheral systems as well as the CNS. Leptin receptors are found in many peripheral tissues, however the short form of the receptor predominates. The signaling efficiency of the truncated forms of the leptin receptor remain to be investigated.

Leptin has been implicated in causing peripheral insulin resistance by inhibiting insulin signal transduction (reviewed by Houseknecht et al., 1998). Leptin treatment of adipocytes reduces insulin stimulation of carbohydrate and lipid metabolism as well as insulin-stimulated protein synthesis. Leptin administration to human hepatic HepG2C cells, rat1 fibroblasts or primary rat adipocytes results in attenuation of insulin-stimulated phosphorylation of IRS-1 or insulin-stimulated glucose metabolism, respectively. Furthermore, leptin receptors have been found on pancreatic beta cells and has been reported to directly inhibit beta-cell secretion of insulin by altering ion channel function. In contrast, studies of leptin action on the muscle C2/C12 cell line demonstrated that leptin stimulates glucose uptake and glycogen synthesis. Additional studies on isolated mouse skeletal muscle have indicated that leptin increases fatty acid oxidation but does not affect insulin-stimulated glucose metabolism.

Leptin may also regulate secretion of cortisol. Leptin is very effective in decreasing bodyweight and food intake when administered to adrenalectomized rats, an effect which is inhibited after supplementation of glucocorticoids. Thus, under normal circumstances, the inhibitory influence of glucocorticoids may prevent hypophagia. In cases of hypercortisolism, which is commonly observed in obese subjects, this mechanism may contribute to leptin resistance.

Leptin and the Genetics of Human Obesity

Monogenic forms of obesity produced in rodent models have been beneficial in leading scientists to identify genes that contribute to obesity. It is estimated that up to 40 to 70 percent of human obesity phenotypes are heritable (Allison, 1996). However, the genetic alterations leading to obesity appear to be much more complex than single-gene mutations, since few cases of single-gene mutations in obese human subjects have been identified. Many individuals may have a combination of genetic alterations which may predispose them to obesity, although environmental factors and behavior also influence the etiology of the obese phenotype.

Human geneticists have used the candidate gene approach to search for potential human obesity genes. Using known genes identified from their obesity effects on animal models, gene linkage analysis has been undertaken by examining numerous candidate genes and various populations of obese human subjects. The statistical support for linkage has been presented in the form of an LOD score (logarithm of the likelihood ratio for linkage) (reviewed by Comuzzie and Allison, 1998). Ideally, a LOD score of 3 would be interpreted as strong evidence for linkage, where this score would represent an hypothesis that genetic linkage is 1000 times more likely than the alternative of no genetic linkage. Currently, there are few candidate obesity genes that have reached this level of significance (Comuzzie and Allison, 1998). There may be two reasons for the low LOD scores. One may be that genes identified in obese animal models may not play an important role in human obesity. A second reason may be that the low statistical score may reflect too small a sample size in the human studies. Also, genetic heterogeneity of human subjects may complicate the statistical studies. In contrast to human obese populations, obese rodent models are derived from closely bred strains, resulting in a more homogeneous genetic background.

Despite the limited findings in linkage between rodent and human obesity genes, the candidate gene approach has yielded some highly significant genetic linkages in human obesity (reviewed by Clement, 1999). An indication of familial genetic linkage was found between the *Ob* locus and severely obese sib-pairs with a with a BMI above 35-40 kg/m^2 in Caucasian-American families. These results suggest that the *Ob* gene region may be implicated in particularly severe forms of obesity in Caucasians, although no linkage was identified in less severe forms of obesity. Studies of Mexican Americans have revealed linkage to a genomic variation in locus D7S514, at a region in close proximity to the leptin gene. D7S514 was implicated in 56% of the variation in skin fold traits within this population. Another nearby locus, D7S495, was linked to variation in BMI, fat mass and waist circumference in this population. In a study of French Canadian families, several genetic variations were located surrounding the same locus. The variations surrounding D7S495 correlated with clinical and metabolic parameters associated with obesity, particularly the amount of subcutaneous fat in subjects. Together, these data suggest that the region surrounding the *Ob* gene or another nearby gene might be important in human obesity in certain populations.

There are some indications that there is a genetic linkage between the *Db* locus (*Ob*-R) and obesity in certain populations. For instance in Pima Indians, an association between obesity and sequence variations in non-coding regions of the *Ob*-R gene were identified. In addition, in French morbidly obese Caucasians, the presence of an insertion/deletion polymorphism in the *Ob*-R was weakly associated with parameters of insulin secretion.

Except in very rare monogenic cases leading to disruption of leptin production in humans, genetic studies have shown that the *Ob* and *Db* loci do not have a major role in common forms of obesity for a majority of the population (reviewed by Clement, 1999). However gene linkage and association studies with these genes suggest that the *Ob* and *Db* genes may have a minor role in the development of obesity. Given the polygenetic nature of human obesity, subtle genetic variations, such as those described that surround the *Ob* and *Db* loci, may serve as susceptibility alleles which could increase the probability that the bearer develops obesity. In this regard, the genetic variation may not be sufficient to drive an onset of obesity, but may interact with other genetic, metabolic or environmental factors to increase an individual's susceptibility of developing obesity.

Leptin Resistance

With the exception of the *ob/ob* mouse and a few documented cases of congenital leptin-deficiency, all current models of rodent and human obesity are characterized by hyperleptinemia and not leptin deficiency. This observation has led to the hypothesis that leptin resistance, defined as a reduced sensitivity to leptin's physiological effects leading to a compensatory increase in plasma levels, occurs in obesity. Although evidence for leptin resistance has been demonstrated in rodents, currently leptin resistance has not been supported by any evidence of physiological insensitivity in human subjects.

Leptin resistance may be caused by a number of potential mechanisms. Analysis of plasma leptin kinetics in humans suggest that the elevated levels observed in obese patients are due to increased leptin production and not to decreased leptin clearance. Over production of leptin may be a compensatory response to faulty signaling of its receptor or down-stream components of the receptor signaling pathway or decreased leptin bioactivity. Superphysiological concentrations of leptin do not trigger maximal leptin effects, suggesting that leptin receptors may become saturated with ligand or that the receptor becomes down-regulated in response to excessive ligand stimulation.

Defects in circulatory transport of leptin may also result in leptin resistance. Studies have demonstrated that leptin circulates specifically bound to at least three proteins in mouse serum and similar evidence exists in humans (reviewed by Houseknecht et al., 1998). In lean mice and humans, most leptin circulates in the bound forms and that the proportion of free leptin is positively correlated with increasing obesity and BMI, indicating that leptin binding proteins are saturated in obese subjects.

The identity of the binding proteins are currently unknown as well as their expression regulation and their definitive interaction with leptin. Many cytokines circulate with soluble forms of their receptor. Multiple isoforms of the leptin receptor exist, including one splice variant which is proposed to be a 85 kDa soluble receptor. In mice, one of the reported leptin binding proteins is also 85 kDa. Immunoprecipitation of leptin from human plasma has demonstrated that only about 10% of free leptin could be immunoprecipitated using a leptin receptor antibody, suggesting that a majority of circulating human leptin is not bound to a soluble receptor.

Leptin is transported to the brain by a saturable system and the efficiency of its transport is reduced in obese patients. Obese subjects with fourfold increases in serum leptin concentrations exhibit only a modest increase in CSF leptin levels (Caro et al., 1996). It is not yet known whether leptin reaches its CNS site of action by direct transport across the blood-brain barrier, by transport from blood to cerebrospinal fluid or by diffusion to sites, such as the hypothalamus, which are functionally outside of the blood-brain barrier. Leptin binding sites have been identified in the choroid plexus and leptomeninges, providing some evidence for a saturable transport system into the brain.

Leptin resistance may not as much reflect a pathological defect in leptin bioactivity but simply reflect its functional limitations in regulating food intake and body fat stores. Spiegelman and Flier proposed that the key evolutionary function of leptin may not be as much as to avoid obesity as it was to function as a metabolic survival factor, where it functions in preventing death by starvation. Thus, leptin could serve as a survival factor to regulate body fat stores during cycles of "feast and famine" (Spiegelman and Flier, 1996).

Clinical Implications for Leptin

In 1997, Amgen, Inc. announced the successful completion of Phase 1 regulatory trials with recombinant leptin. Recombinant leptin was administered by injection and a safe dose range for humans was established, however there were injection site reactions, particularly with multiple daily injections at the higher doses (Business Wire, June 17, 1997). The final report on the conclusion of the first clinical trial was disappointing. A considerable number of patients withdrew from the study because of problems with skin irritation and swelling at the injection sites and other patients were dropped from the study because of their lack of compliance to their assigned dietary regimen. Significant weight loss was only observed in the group of patients receiving the highest doses of leptin. Of the 47 patients who completed the study, the eight receiving the highest dose lost an average of 7.1 kg, while the twelve taking a placebo lost 1.3 kg. There was a wide variation of effects among the patients in the group treated with the highest dose, from a loss of about 15 kg to a gain of 5 kg (Heymsfield et al., 1999).

Based upon the clinical studies, the use of recombinant leptin has some drawbacks, however, considering leptin as a target for future antiobesity drugs does remain promising. One drawback for using recombinant leptin is that administration must occur by injection because oral administration of leptin would result in its breakdown in the gastrointestinal tract. It is also projected that the doses required for achieving satisfying weight reduction would be high as well as expensive. Alternative small molecule strategies many provide additional promise for developing successful antiobesity therapeutics. For instance, the relationship of leptin to human obesity indicates a defective response to leptin in contrast to a lack of leptin production. Developing molecular analogues to leptin which would either more effectively stimulate the leptin receptor or perhaps have greater effectiveness in being transported to the receptors in the hypothalamus, may result in an efficacious anti-obesity therapeutic.

Other components of the leptin pathway may also serve as promising therapeutic targets. For instance, beta3-adrenergic receptors may mediate afferent signals in the leptin feedback loop system, since agonist stimulation of the beta3-adrenoceptor for 24 hours resulted in both decreased leptin mRNA expression and a 20% reduction in circulating leptin levels, whereas levels of NPY and MCH remained unchanged (reviewed by Lonnqvist et al., 1999). These results suggest that the beta3-adrenoceptor functions in the leptin pathway downstream of leptin, NPY and MCH. It is well known that stimulation of beta3-adrenergic receptors in rodents results in weight reduction due to increased thermogenesis and decreased fat depots. In addition, beta3 agonist stimulation of mice on a high fat dietary regimen resulted in normalization of leptin mRNA levels and prevented obesity, but not hyperphagia. A recent breakthrough in characterizing the leptin signaling pathway includes the discovery of suppressor of cytokine signaling-3 (SOCS-3). Leptin appears to stimulate production of SOCS-3, which in turn feeds back to inhibit further leptin signaling. It is possible that leptin resistance observed in obese patients may be due to SOCS-3 overexpression and/or hyperactivity (Chicurel, 2000). Thus, generation of SOCS-3 inhibitors currently appear to be an exciting new target in the development of antiobesity drugs.

Characterization of the leptin signaling pathway has also led to discovery of weight-regulating properties of other gene products. For instance, ciliary neurotrophic factor (CNTF) is a cell surface receptor that closely resembles the leptin receptor, is expressed in similar areas of the hypothalamus and activates similar intracellular signaling pathways. In clinical trials using CNTF to treat amyotrophic lateral sclerosis, CNTF administration unfortunately did not show much neurological improvement, however the patients did exhibit significant weight reduction. Experiments have also been conducted using recombinant CNTF or derivatives of CNTF in both obese mouse mutants and in mice made obese by high fat dietary regimens. Administration of these molecules resulted in considerable loss of fat depots and overall weight, while maintaining moderate eating habits (reviewed by Chicurel). Currently, Regeneron is conducting clinical trials of the CNTF derivative, axokine. High doses of the drug have been reported to cause cold sores and nausea, however patients have tolerated the low doses well. Most patients lost significant amounts of weight, reducing food intake by about 500 calories a day (reviewed by Chicurel).

Much remains to be characterized in understanding downstream effectors of leptin in the afferent pathways by which the hypothalamus regulates food intake and energy expenditure. Also, refined understanding of how leptin affects peripheral target organs, such as the liver, pancreas and reproductive system, need to be accomplished. With the help of gene-function studies in rodent models, these pathways are likely to become better characterized, enabling additional genes to be identified that could serve as promising targets for developing antiobesity drug therapeutics.

REFERENCES

Ahima RS, Prabakaran D, Mantzoros C, Daqing QU, Lowell B, Maratos Flier E *et al.* Role of leptin in the neuroendocrine response to fasting. *Nature* 1996; 382:250-252.

Allison DB. The use of discordant sibling pairs for finding genetic loci linked to obesity: Practical considerations. Int *J Obes* 1996; 20:553-560.

Augustine KA, Rossi, RM. Rodent mutant models of obesity and their correlations to human obesity. *Anat Rec (New Anat)* 1999; 257:64-72.

Bouchard C, Perusse L, Genetics of obesity. *Annu Rev Nutr* 1993; 13:337-354.

Caro JF, Kolaczynski JW, Nyce MR, Ohannesian JP, Opentatova I, Goldman WH *et al.* Decreased cerebrospinal-fluid/serum leptin ratio in obesity: a possible mechanism for leptin resistance. *Lancet* 1996; 348:159-161.

Chicurel M. Whatever happened to leptin? *Nature* 2000; 404:538-540.

Clement K. Leptin and the genetics of obesity. *Acta Paediatr Suppl* 1999; 428:51-57.

Coleman DL. Effects of parabiosis of obese with diabetes and normal mice. *Diabetologia* 1973; 9:294-298.

Comuzzie AG, Allison DB. The search for human obesity genes. *Science* 1998; 280:1374-1377.

Considine RV, Sinha M, Heiman M, Kriauciunas A, Stephens T, Nyce M, Ohannesian J, Marco C. 1996. Serum immunoreactive-leptin concentrations in normal-weight and obese humans. *N Engl J Med* 1996; 334:292-295.

Echwald SM. Genetics of human obesity: lessons from mouse models and candidate genes. *J Intern Med* 1999; 245:653-666.

Gwo-Hwa L, Proenca R, Montez JM, Caroll KM, Darvishzadeh JG, Lee JI *et al.* Abnormal splicing of the leptin receptor in diabetic mice. *Nature* 1996; 379: 632-635.

Haalas JL, Gajiwala KS, Maffei M, Cohen SL, Chait BT, Rabinowitz D *et al.* Weight-reduction effects of the plasma protein encoded by the obese gene. *Science* 1995; 269:543-546.

Heymsfield SB, Greenberg AS, Fujioka K, Dixon RM, Kushner R, Hunt T, Lubina JA, Patane J, Self B, Hunt P, McCamish M. Recombinant leptin for weight loss in obese and lean adults: A randomized, controlled, dose-escalation trial. *JAMA* 1999; 282: 1568-1575.

Houseknecht KL, Baile CA, Matteri RL, Spurlock ME. The biology of leptin: A review. *J Anim Sci* 1998; 76:1405-1420.

Hummel KP, Dickie MM, Coleman DL. Diabetes, a new mutation in the mouse. *Science* 1966; 153:1127-1128.

Ingalls AM, Dickie MM, Snell GD. Obese, a new mutation in the mouse. *J Hered* 1950; 41:317-318.

Kolaczynski JW, Ohannesian JP, Considine RV, Marco CC, Caro JF. Response of leptin to short-term and prolonged overfeeding in humans. *J Clin Endocrinol Metab* 1996b; 81:4162-4165.

Lonnqvist F, Nordfors L, Schalling M. Leptin and its potential role in human obesity. *J Intn Med* 1999; 245:643-652.

Montague CT, Farooqui JP, Whitehead MA, Soos H, Rau NJ, Wareham CP, Sewter JE *et al.* Congenital leptin deficiency is associated with severe early-onset obesity in humans. *Nature* 1997; 387:903-908.

Pelleymounter MA, Cullen MJ, Baker MB, Hecht R, Winters D, Boone T *et al.* Effects of the obese gene product on body weight regulation in ob/ob mice. *Science* 1995; 269:540-543.

Seeley RJ, Yagaloff KA, Fisher SL *et al.* Melanocortin receptors in leptin effects. *Nature* 1997; 390:349.

Spiegelman BM and Flier JS. Adipogenesis and obesity: Rounding out the big picture. *Cell* 1996; 87:377-389.

Strobel A, Issad T, Camoin L, Ozata M, Strosberg AD. A leptin missense mutation associated with hypogonadism and morbid obesity. *Nature Genetics* 1998; 18:213-215.

Stunkard AJ, Foch TT, Hrubec Z. A twin study of human obesity. *JAMA* 1986; 256:1213-1215.

Vogler GP, Sorenson TIA, Stunkard AJ, Srinivasan MR, Rao DC. Influences of genes and shared family environment on adult body mass index assessed in an adoption study by a comprehensive path model. *Int J Obesity* 1995; 19:40-45.

Zhang Y, Proenca R, Maffei M, Barone M, Lori L, Friedman JM. Positional cloning of the mouse obese gene and its human homologue. *Nature* 1994; 372:425-432.

Chapter 8

GENETIC SUSCEPTIBILITY OF RODENTS TO DIET-INDUCED OBESITY

Joanne Harrold
University of Liverpool

Abstract Obesity results when food intake exceeds energy expenditure. It is a complex multifactoral disease that appears to involve the interaction of numerous neuronal circuits. The current theory states that phenotypic expression of obesity depends on the influence of environmental factors upon an underlying pool of genes that determine susceptibility. However, the nature of this inherited predisposition remains unclear. Recent years have witnessed considerable progress in our understanding. Rodents continue to be central to this study, providing ideal models to determine the means by which the various regulatory pathways interact in different feeding states. In particular, observations from the dietary-induced obese state have yielded valuable information regarding the identity of candidate susceptibility genes. Mutations within some of the lead players identified have since been discovered in obese humans. Such information is a pre-requisite for successful treatment of the disorder and the curbing of further increases in body weight across future generations.

INTRODUCTION

Obesity is an increasingly costly and important health problem, which has reached epidemic proportions in the developed countries of the world, particularly in areas that have recently changed from traditional to Westernised lifestyles (Mokdad *et al.*, 1999). A clear example is the Pima Indians who now demonstrate a high incidence of obesity, following their exposure to American diet and culture (Reid *et al.*, 1971).

This increasing prevalence is a strong indication that environmental factors are an important cause of the epidemic. At first glance the management of excess body weight may therefore appear obvious namely, the combination of reduced food consumption and increased physical

J.B. Owen et al. (eds.), Animal Models – Disorders of Eating Behaviour and Body Composition, 133–155.

activity. However, even with such modifications in behaviour, reductions in body weight are resisted by very robust physiological mechanisms, implying that other factors play a significant contribution. Recent advances in obesity research, specifically results from family-based studies, strongly suggest that a hereditary component also plays a substantial role (Bouchard and Tremblay, 1997). Various authorities have estimated this genetic component to be responsible for 40-70% of the variation in obesity-related conditions in humans (Comuzzie and Allison, 1998); its genetic basis is in most cases highly complex, with obesity being inherited as a polygenic trait involving a number of genes operating simultaneously.

Apart from rare obesity-associated syndromes, genetic influences on body weight appear to operate via 'susceptibility genes' - i.e. those that exacerbate the risk of developing a condition but are not themselves essential for its occurrence or alone cannot account for its development; differences in genetic susceptibility within a population therefore determine those individuals most likely to become obese under any given set of environmental parameters. This overall genetic susceptibility may partly relate to the survival pressures that have operated throughout the vast majority of human evolution. The main threat to survival has been the shortage of food; mammals have therefore evolved sensitive and sophisticated physiological mechanisms to defend against weight loss during energy deficit, but have weaker responses to prevent weight gain when food is abundant. This prinicple is referred to as 'The Thrifty Genotype Hypothesis' (Neel, 1962).

The phenotypic expression of the thrifty genotype ultimately depends on the influence of environmental factors to unmask tendencies to develop obesity (Figure 1). In animals genetically predisposed to obesity, numerous abnormalities of neural function have been identified that encourage them to become obese if the energy content, palatability and/or availability of diet increase. Once obesity has developed, the abnormally increased body weight is then defended against weight loss; effectively, obesity then becomes the norm for such individuals. Various mechanisms have been proposed for this apparent increase in the 'set point' around which body fat mass is regulated. One is 'metabolic imprinting' – the physical modification of existing neuronal systems or the formation of new neural networks that regulate energy homeostasis. This plasticity could theoretically be passed onto future generations. Such a feed-forward cycle would perpetuate body weight gain across generations. Identification of the obesity susceptibility genes may therefore help to prevent further escalations in human obesity and its associated health complications.

The scope of this chapter is to summarise our current understanding of the molecular mechanisms of normal body-weight regulation and the manner in which these pathways can be altered early in life, leading to obesity, or alternatively become defective in the obese state. Mammalian body-weight regulation is extremely complex, with food intake and energy expenditure depending on numerous networks of regulatory neuronal pathways. Genetic approaches have been critical in identifying the key components of this system and in identifying the mutations which predispose individuals to obesity.

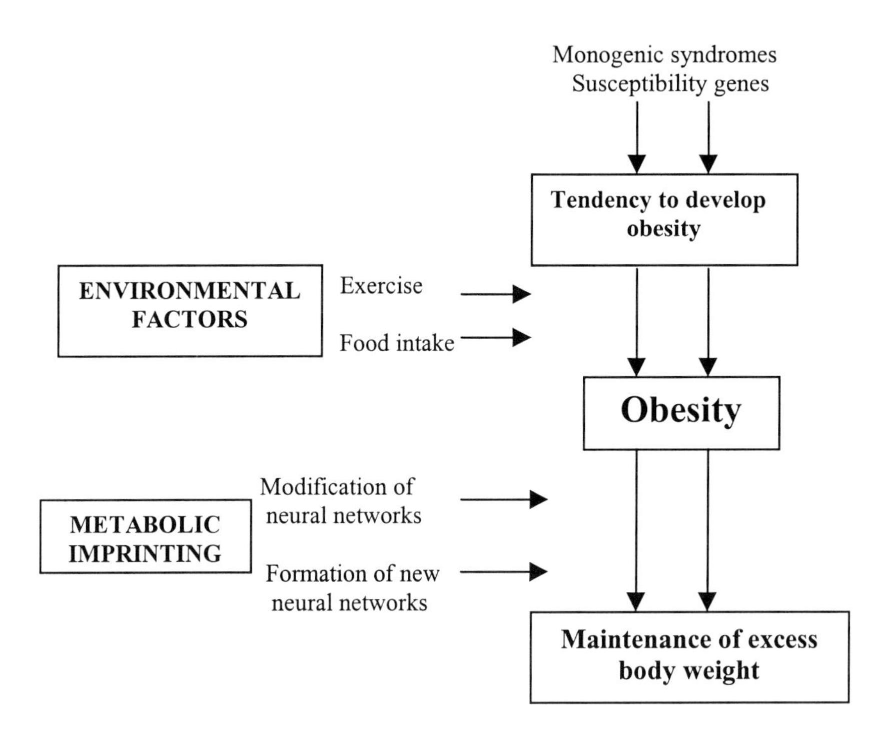

Figure 1 **Factors influencing the development of obesity**

INFLUENCES OF MATERNAL ENVIRONMENT AND GENETIC BACKGROUND

Several maternal conditions have been linked to the development of obesity and diabetes in both human and rodent offspring (Berenson *et al.*, 1997; Perseghin *et al.*, 1997). As both of these conditions can be inherited it has been difficult to determine whether genetic background or adverse intrauterine and postnatal environments are the cause of this phenomenon. Various animal models have been used to shed light on this issue, with results indicating that both factors must be present to increase the likelihood of the development of obesity and diabetes in progeny.

In rats, mothers with either type I (Silverman *et al.*, 1995) or type II diabetes (Perseghin *et al.*, 1997) tend towards obese offspring. Additionally, both maternal food deprivation (Jones *et al.*, 1986) and obesity (Levin and Govek, 1998) result in obese progeny. Some of these conflicting findings have been explained through the study of substrains of rats bred for their propensity to either resist or develop obesity when exposed to a high-fat diet (Levin *et al.*, 1997). They demonstrate that the association of such environmental factors with excess body weight is dependent on the offspring being genetically predisposed to become obese (Levin *et al.*, 1998), thus verifying that genotype is a critical independent factor in determining outcome.

Hypothalamic neuropeptides are central to the regulation of body weight and food intake. Changes in specific examples of neuronal systems, namely galanin and neuropeptide Y (NPY), have been identified in neonatal rats exposed to various maternal perturbations of energy balance. Increased numbers of galanin and NPY neurons have been reported in neonatally overfed rats (Plagemann *et al.*, 1999a, b). Abnormal increases in both NPY concentration and neuron number have also been identified in neonatally hyperinsulinaemic offspring of diabetic mothers (Plagemann *et al.*, 1999c). Both NPY and galanin stimulate food intake and their overactivity could therefore provide the underlying pathophysiological mechanisms for the development of obesity. Polymorphisms that affect the expression or release of these or other appetite-modulating neurotransmitters could thus, in theory, represent potential susceptibility genes.

Insulin - a classical signal of nutritional status – also appears to play a critical role in the development of obesity in neonates. Exogenous insulin administration in high doses, produces obesity in offspring, and this is associated with altered noradrenergic signalling (Perseghin *et al.*, 1997). Direct insulin injection into the hypothalamic nuclei of pups during the first 2 postnatal weeks has the same consequences (Plagemann *et al.*, 1992). Such abnormally elevated insulin levels may be responsible for exposing the dysfunction of regulatory neuropeptide systems. Their overactivity might represent an acquired resistance to increased levels of the satiety factor (Plagemann *et al.*, 1999b).

The timing of changes in the maternal condition in respect to the development of the nervous system also appears to play a role in inducing and maintaining obesity in predisposed individuals. Elevated insulin levels during gestation and the early postnatal period are known to produce progeny with abnormal neuronal size and number of hypothalamic structures implicated in the regulation of body weight and metabolism (Plagemann *et al.*, 1999d). Insulin is a neuronal growth factor (Folli *et al.*, 1996), influencing neuronal differentiation (Nataf and Monier, 1992), survival (Tanaka *et al.*, 1995), neurite development (Yamaguchi *et al.*, 1995), synaptic transmission (Puro and Agardh, 1984) and the density of α_2-adrenoreceptors (Levin *et al.*, 1998). Exposure to high insulin levels during the development of the brain may permanently alter the pattern of neural innervation. Such metabolic imprinting may in turn promote the maintenance of excessive body weight introduced by dysfunction of key hypothalamic neuronal systems. However, it has not been convincingly proven whether these changes are the cause or effect of the obesity that accompanies them.

Besides raised insulin, numerous other maternal metabolic abnormalities can promote obesity in offspring through the unmasking of genetic susceptibility and also potentially through metabolic imprinting on genetically susceptible neural circuits during development. The influences of the numerous variables identified in young animals apparently have long lasting effects on the systems controlling body weight, being linked to increased body weight and fat content in adult life (Faust *et al.*, 1980). Alterations to hypothalamic neuronal systems have also been found to persist into adult life, with elevated numbers of galanin and NPY neurons identified in adult animals from gestationally diabetic mothers (Plagemann *et al.*, 1999c). However, whether such abnormalities are transmitted to the offspring of these affected animals remains to be determined.

ADULT DIETARY OBESITY

Very little is understood regarding the factors predisposing humans to obesity. This is mainly because it is extremely difficulty to identify obesity-prone individuals before they actually develop obesity and its associated metabolic abnormalities. Studies of apparently obesity-prone populations are beginning to reveal some potential discriminatory factors. For example, young Pima Indians whose energy expenditure lies in the lowest quartile appear to be at the highest risk of developing obesity (Spraul *et al.*, 1993). Yet, such predictors remain very limited, and any links between obesity and the neuronal dysfunction identified in obese rodents remain entirely speculative, as the study of functional anatomy and regulatory neuropeptides in the human brain is still in its infancy (Gautier *et al.*, 1999).

Animal models therefore continue to provide surrogate systems for expanding our understanding. Models of monogenic obesity syndromes e.g. *fa/fa* Zucker rats and *ob/ob* and *db/db* mice, have provided some useful insights into the regulatory mechanisms controlling body weight and food intake. However, there are fundamental problems associated with the use of such models – particularly the fact that analagous mutations within the human obese population are vanishingly rare. By contrast, the rodent models of dietary-induced obesity (DIO) provide an alternative and potentially helpful means of studying obesity-prone and obesity-resistant individuals.

Rodent model of dietary-induced obesity

It has long been known that different strains of rat differ markedly in their susceptibility to obesity when exposed to an energy-dense palatable diet. Interestingly, such differences also occur within the same strain of rodent: roughly one-half of unselected rats or mice become obese when placed on an energy-dense diet (Hill *et al.*, 1983; Levin *et al.*, 1983; Harrold *et al.*, 2000), whilst the remainder appear resistant to the development of obesity. Relatively little is know about how such variability arises among animals from an apparently genetically similar background, although recent work suggests that this may be explained by genotypically-distinct populations of animals occurring within a defined strain.

Studies of DIO have shown that obesity-prone and obesity-resistant rats can be identified prospectively before expressing their respective phenotpyes (Levin and Sullivan, 1989a; Levin, 1995; Harrold *et al.*, 2000). Additionally, the selective breeding of substrains of dietary-susceptible and dietary-resistant rats has enabled manipulation of their prenatal environment, thus allowing specific neural and metabolic functions to be altered (Levin *et al.*, 1997). Using these methods, some factors that predispose rats to become obese have been identified. DIO is evidently inherited as a polygenic trait in a similar manner to that of human obesity, candidate genes that determine obesity and its metabolic may affect both sides of the energy-balance equation, namely appetite regulation and energy expenditure (Figure 2).

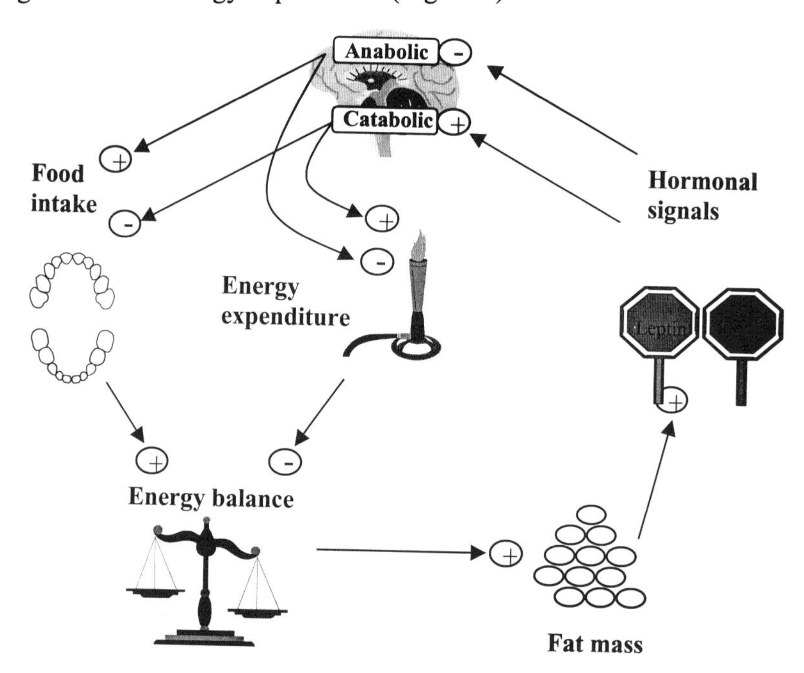

Figure 2. Model indicating potential sources of susceptibility genes within the network of pathways regulating body weight. (Adapted from Schwartz *et al.*, 2000)

Sympathetic Nervous System Activity

Both monogenetic and diet-induced obese rats are frequently hyperinsulinaemic and/or hyperglycaemic (Zucker and Antoniades, 1972; Levin *et al.*, 1981). Moreover, defects in thermogenesis have been identified in some obese rodents (Godbole *et al.*, 1978; Levin *et al.*, 1981). As insulin release, glucose metabolism and thermogenesis are all modulated by inputs from the sympathetic nervous system (SNS; Wright and Malaisse, 1968; Depocas *et al.*, 1978; Foster and Frydman, 1978), it has been proposed that differences in SNS activity may play a role in the interstrain variability in susceptibility to DIO.

Levin and colleagues have widely studied the role of SNS in dietary-susceptible and dietary-resistant rats (Levin *et al.*, 1986; Levin *et al.*, 1987; Levin and Sullivan, 1987 and 1989). They found that it was possible to predict which rats were susceptible to DIO by measuring activiation of the SNS to either intravenous (Levin *et al.*, 1993 and 1986) or intracarotid (Levin *et al.*, 1992) glucose infusion. Susceptible rats demonstrated an exaggerated activation, as determined by plasma levels of noradrenaline. The animals were also found to have higher levels of 24 hr urinary noradrenaline (Levin and Planas, 1993). A robust reciprocal relationship has been identified between food intake and SNS activity, in which increased SNS activity produces satiety, increased thermogenesis and decreased body fat (Bray, 1991a and 1991b). These results would therefore suggest that dietary-susceptible rats were attempting to compensate for excesses in calorie-intake by increasing sympathetic function, particularly in thermogenically active tissue. However, using these discriminators, it was found that dietary-susceptible rats actually have reduced sympathetic activity in the heart and pancreas, but not in brown adipose tissue (an important thermogenic tissue in the rat). In light of the relationship between food intake and SNS activity, this low activity would be expected to lead to hyperphagia, and in the absence of other compensatory systems, produce obesity.

Studies comparing Sprague Dawley and Fischer F344 rats have further highlighted the importance of decreased pancreatic sympathetic activity and its influences on insulin levels in the development of rodent obesity. Fischer F344 rats eat less and become less obese than Sprague Dawley and their pancreatic noradrenaline remains comparable to that in lean controls (Levin *et al.*, 1983). By contrast, decreased pancreatic noradrenaline turnover associated with hyperinsulinaemia is characteristic of DIO Sprague Dawley rats (Levin *et al.*, 1987). As the sympathetic input to the pancreas inhibits insulin release in rodents (Campfield and Smith, 1980), the hyperinsulinaemia of DIO rats may therefore result, at least in part, from decreased sympathetic inhibition of insulin release. Such persistent hyperinsulinaemia could, in turn, impede decreases in food intake and metabolic efficiency, possibly through the development of resistance to elevated levels of this satiety factor. Therefore, distinct differences in the regulation of pancreatic sympathetic activity and plasma insulin appear to have a permissive effect on the development and maintenance of the obese state and as such represent potential susceptibility genes.

However, decreased pancreatic sympathetic activity and hyperinsulinaemia do not appear to be features common to all obese rodents and are thus not prerequisites for the development of DIO. It has been found that young Sprague Dawley rats develop obesity despite increased NE turnover and normal plasma insulin levels (Levin *et al.*, 1987). The identified changes may therefore simply be a consequence of chronic obesity and reflect the age of the animal.

Leptin

A key component of the physiological systems regulating appetite and body weight is the hormone leptin, which is produced by adipose tissue and circulates in proportion to body adiposity. It is postulated to provide information on the nutritional status of the body to regulatory centres, primarily in the hypothalamus of the brain. An increase in body fat is associated with increased levels of leptin and would be predicted to result in reduced food intake, whilst a decrease in body fat leads to lower leptin levels and the opposite effects (Friedman and Halaas, 1998). This tightly regulated endocrine feedback loop may help to explain why body weight (and fat content) remain remarkably constant in normal rodents despite considerable daily variations in energy intake and expenditure during adult life. Moreover, a break down in this self-regulation may underlie the development of obesity in DIO rodents, although its significance in humans remains uncertain.

Evidence suggests that leptin secretion can become transiently dissociated from levels of body fat. For example, fasting lowers plasma leptin concentrations in both rodents and humans more rapidly and to a greater extent than would be predicted from the decrease in body fat (Schwartz *et al.*, 2000). Such an exaggerated early decline would potentially enable compensatory responses to come into play before energy stores are severely depleted. Similarly, an exaggerated rise in plasma leptin occurs in some rats soon after exposure to a palatable diet (Harrold *et al.*, 2000a). We have now found that this rise in leptin can be negatively correlated to the degree of obesity developed several weeks later during continued exposure to the diet (Harrold *et al.*, 2000a) – perhaps this offers another potential measure to predict which rats are susceptible to DIO. Our findings are consistent with the conclusion that lower plasma leptin levels early in adult life, were associated with greater subsequent weight gain in a population of Pima Indians (Ravussin *et al.*, 1997) and also C57BL/6J mice (Surwit *et al.*, 1997). The leptin system may therefore exert some influences across various mammalian species.

Recent research suggests that leptin itself, popularly hailed as a 'fat buster', may actually be a 'thrifty' hormone functioning mainly as a regulator of the body's response to starvation in order to protect energy stores. Leptin injection into rats on a low-calorie diet has been found to have little influence on leptin mRNA levels in fat cells, but induces leptin expression in muscle cells, where the gene is not normally expressed (Wang *et al.*, 1998). This novel site of expression is assumed to represent a signal for the muscle to utilise fat rather than deplete protein and carbohydrate stores when these cannot readily be replaced. However, leptin injection into animals fed an energy-dense diet causes a dramatic fall in leptin mRNA in fat cells, but little change in muscle expression. This indicates that leptin curbs its own production in fat cells, possibly to ensure that the animals will eventually eat again and restore fat depots. In this situation, the muscle will also utlise less fat than in animals fed a low-fat diet. Constant overfeeding could therefore limit the effectiveness of leptin's attempts to maintain body weight within a narrow range, resulting in DIO.

It remains to be seen whether the responses of leptin in fat and muscle cells vary between dietary-susceptible and dietary-resistant rodents. However, the case for the *obese* gene, which encodes leptin, being applicable to the common human obesity problem, has been somewhat weakened by several lines of evidence. Surwit and colleagues (2000) attempted to bypass the low initial leptin response observed in C57BL/6J mice (a well characterised model of DIO and diabetes), following high-fat feeding. They found that leptin supplementation slowed the development of DIO but did not prevent it. Additionally, while mutations in the *obese* gene lead to hyperphagia and severe obesity in animals (Bray and York, 1979) defects in this gene are vanishingly rare in human obesity (Montague *et al.*, 1997). However, recombinant leptin therapy in a child with congenital leptin deficiency has been shown to successfully lower appetite (which is greatly increased and thus the main cause of obesity) and mobilise body fat (Farooqi *et al.*, 1999). Therefore, despite the scarcity of these cases they do confirm the overall importance of the leptin signal in the regulation of body weight.

Interestingly, 5-10% of obese human subjects have relatively low leptin levels in the absence of mutations in the *obese* gene (Maffei *et al.*, 1995) and may therefore represent a state of relative leptin deficiency. It is not yet clear how much of this is genetically determined. Efforts to define regulatory elements within the leptin promotor and other pathways that may limit leptin gene expression in such individuals have identified potential sequences for mutational screening. Analysis of these may shed more light on the genetic susceptibility of this minor subpopulation. One candidate gene is glucosamine, the end-product of glucose flux via the hexosamine biosynthetic pathway, high intracellular levels of which up-regulate leptin in fat and muscle (Wang *et al.*, 1998).

Does leptin resistance exist?

As leptin is produced in proportion to body adiposity, most human obesity is associated with increased rather than decreased levels of leptin in the blood stream (Maffei *et al.*, 1995). The single-gene mouse mutants *Agouti*, *tubby* and *fat* (Frederich *et al.*, 1995), in addition to rodents with established DIO (Harrold *et al.*, 2000a) also have very high levels of circulating leptin. For some reason, these individuals simply fail to respond to the hormonal signal. This led to the theory that, similar to type 2 diabetes, obesity involves resistance to the action of a hormone (Hamilton *et al.*, 1995). One potential mechanism of resistance is impaired leptin transport across the blood-brain barrier (BBB); this transport potentially occurs by one of the short forms of the leptin receptor (Ob-Ra), which lacks the cytoplasmic signalling domain of Ob-Rb (the predominant signalling form of the receptor). Studies examining CSF leptin levels following intravenous bolus injection of the hormone (Burguera *et al.*, 2000) or in extremely obese individuals with high plasma leptin (Caro *et al.*, 1996) found that CSF levels are disproportionately low in obese subjects in comparison to their plasma levels. This may suggest that inward leptin transport, and perhaps Ob-Ra in the BBB are easily saturated.

Once in the brain, leptin's actions are mediated by interaction with the long form of the receptor Ob-Rb, expressed at high levels in the hypothalamus. Interestingly, recent work implies that down regulation of Ob-Rb protein levels in the hypothalamus may also be associated with hypothalamic leptin resistance in DIO rats fed a high-fat diet (Madiehe *et al.*, 2000). Alternatively, leptin resistance may stem from molecules downstream of the leptin receptor that interfere with its signalling. One example is SOCS-3 (a supressor of cytokine signalling-3), whose expression is activated by leptin binding and STAT-3 activation, perhaps in an autoregulating fashion. It has been found that injection of leptin increases production of SOCS-3 in hypothalamic cells bearing the Ob-Rb (Bjorbaek *et al.*, 1998). This in turn may prevent further signalling by the receptor. Mutations in SOCS-3, which is presumably a component of the normal mechanism for halting leptin signalling, could in theory explain resistance to leptin and lead to weight gain by interfering with the negative feedback in which leptin acts on the CNS to prevent increases in body fat mass.

As there are numerous hypothalamic responses to leptin there may be multiple sites of leptin resistance. Identifying relevant genes is critical to determining their potential roles in predisposing to obesity. IT could be argued that leptin and its receptors are ultimately poor candidates to be thrifty genes as their mutations not only result in weight gain but also cause infertility, and thus prevent positive selection. Additionally, while loss of function mutations in leptin receptors lead to obesity in monogenic rodent models e.g. *db/db* mice (Chen *et al.*, 1996), leptin resistance in obese individuals is very rarely attributed to mutations in the leptin receptor. Consequently research efforts are now directed downstream of leptin function in an attempt to identify candidate genes and thus potential targets for anti-obesity treatment.

Hypothalamic Neuropeptide Effectors of Leptin Signalling

The role of leptin in the regulation of energy homeostasis is unquestionable. Various neuronal populations appear to be capable of integrating the circulating leptin signal from the periphery and forwarding the information to effector systems so as to control energy homeostasis; these neuronal systems are therefore potential sites harbouring susceptibility genes. A number of abnormalities in the function of such systems have been identified in dietary-susceptible rats, many of which are normalised once obesity fully develops, possibly through metabolic imprinting on genetically altered circuits. The most likely sites are integrator-effector neurons in the hypothalamic arcuate nucleus (ARC), which monitor a host of signals from the periphery and surrounding brain regions that carry information regarding the nutritional status of the body. Here, leptin's regulation of energy balance is mediated via numerous neuropeptides. Catabolic neuropeptides, the actions of which are stimulated by leptin, act to decrease food intake and increase energy expenditure, whilst anabolic peptides are inhibited by leptin and have the opposite effects. Determining which of these numerous molecules are critical in terms of body weight regulation is essential to the identification of potential susceptibility genes. Amongst the mediators of leptin's actions, two present particularly promising targets; the anabolic peptide neuropeptide Y (NPY) and the catabolic melanocortins.

Neuropeptide Y

The 36 amino-acid neurotransmitter NPY is the dominant anabolic neuropeptide in the hypothalamus. Central NPY injection potently stimulates food intake in rodents and repeated injections result in obesity (Stanley *et al.*, 1986). NPY injection has also been found to reduce oxidation of dietary fat, a characteristic common to the development of DIO in rats (Chang *et al.*, 1990) and humans (Rising *et al.*, 1996). Theoretically, this should favour increased fat deposition upon ingestion of high-fat food. The ARC NPY neurons are mainly responsible for these reported effects, with overexpression of ARC NPY mRNA identified in a number of models of genetic hyperphagia, including obesity-prone Zucker rats (Sanacora *et al.*, 1996) and *ob/ob* mice (Wilding *et al.*, 1993), as well as following experimental food deprivation (Sahu *et al.*, 1988). Alterations to the NPY system have also been identified in hyperinsulinaemic and over-fed neonates. These include increased NPY-dependent inhibition of the ventromedial hypothalamic nucleus (VMH) satiety signal (Heidel *et al.*, 1999) and potential leptin or insulin resistance of the NPY system (Plagemann *et al.*, 1999b). Such early variations in rodents have been found to have long term influences, being associated with obesity in adult hood (Faust, 1980).

Other specific components of the NPY system where mutations could influence susceptibility to DIO include the NPY receptors. Currently, six different subtypes have been characterised, with evidence for further NPY receptor subtypes (O'Shea *et al.*, 1997; Balasubramaniam, 1997). Only five have been characterised in rat brain, namely Y_1-Y_5. Originally, it was concluded that Y_1 solely mediates the stimulatory effects of NPY on food intake through increased carbohydrate intake and meal size (Leibowitz *et al.*, 1991). However, more recent work has implicated Y_5 as the feeding receptor; knockout of the Y_5 receptor or administration of an antagonist being able to inhibit both NPY-induced hyperphagia and the hyperphagia present in genetically obese models (Schaffhauser *et al.*, 1997). However, there is still controversy as to which of the NPY receptors represents the true feeding receptor, and both Y_1 and Y_5 are still regarded as potential targets for novel anti-obesity drugs.

However, the importance of an overactive NPY system in DIO remains speculative. The finding that mice lacking NPY have intact feeding responses and grow normally, raises questions about the importance of the role of NPY in the development and determination of susceptibility to DIO. Yet, this does not necessarily disprove the importance of the NPY system in normal food intake regulation, but highlights the potential for other neuronal systems to take over the leading role of NPY and compensate for its deficiency. Further doubts surround the interpretation of the observed activity of the NPY system in rodents with DIO. In such animals, both NPY mRNA concentration and receptor activity are decreased (Widdowson *et al.*, 1997). One interpretation is based on the theory that the obese state is proposed to be that which must be achieved in DIO animals before an altered NPY system can operate normally. Consequently the reduced activity reflects a correction of the original overexpression and dysregulation required to develop obesity. This explanation is supported by evidence that DIO rodents robustly defend their elevated body weight against food deprivation through elevation of NPY mRNA (Levin and Dunn-Meyell, 2000). According to this hypothesis, genetically programmed NPY overexpression in dietary-susceptible rats will be responsible for making them susceptible to obesity when fed an energy-dense diet and for maintaining the elevated body weight should it be challenged by a change in nutritional status. An alternative interpretation suggests that the observed changes in the NPY system in DIO animals actually represent inhibition of the NPY neurons in an attempt to limit over-eating: NPY would not mediate the hyperphagia that leads to obesity. Instead, the NPY system may operate to protect animals during periods of inadequate food supply by increasing the drive to eat and thus restore energy stores; breakdown of other elements of the energy balance homeostatic system would determine susceptibility to DIO.

The Melanocortins

The melanocortin peptides, of which the archetype is α-melanocyte stimulating hormone (α-MSH) are cleaved from a common precursor molecule pro-opiomelanocortin (POMC). They exert their effects through binding to a family of five receptors, of which both MC3-R and MC4-R are expressed in the hypothalamus (Mountjoy *et al.*, 1994). The observations that central injection of MC4-R selective synthetic receptor agonists suppresses food intake, whilst MC4-R antagonists or MC4-R knockout in mice stimulate feeding and cause obesity (Fan *et al.*, 1997; Huszar *et al., 1997*), further support the notion that tonic signalling by MC4-R limits food intake.

Additional evidence for the importance of melanocortin signalling, in particular through MC4-R, in energy homeostasis comes from studies of the monogenic mouse model of obesity, the yellow agouti mouse (A^y/a). Mutation of the agouti gene in these rodents leads to ectopic expression of the peptide and results in antagonism of hypothalamic MC4-R (Lu *et al.*, 1994). More recently, a novel hypothalamic peptide termed agouti gene-related peptide (AGRP) has been cloned, based on its homology to agouti (Ollmann *et al.*, 1997). AGRP is an endogenous antagonist at both MC3-R and MC4-R. The hypothalamic expression of this peptide, like that of NPY and POMC, is localized to the ARC (Shutter *et al.*, 1997) and is up-regulated by fasting (Broberger *et al.*, 1998) and by leptin deficiency in the *ob/ob* mouse (Shutter *et al.*, 1997), indicating that tonic antagonism of CNS melanocortin receptors is an important factor in the control of body weight. Consistent with this is the hyperphagia observed upon central injection of AGRP (Rossi *et al.*, 1998). These effects are sustained, lasting longer than NPY-induced stimulation of food intake. Consequently, AGRP is currently an area of considerable interest in terms of susceptibility to DIO.

Recent work has confirmed that hypothalamic signalling through MC4-R, and in particular the relationship between this and leptin levels, contributes to intrastrain variability in susceptibility to DIO (Harrold *et al.*, 2000a). It was found that an early rise in plasma leptin, soon after exposure to palatable food and preceding weight gain, predicted a lesser weight gain some weeks later. Leptin may therefore programme the hypothalamus in some way to enable an individual rat to resist over-eating in the long term. Intriguingly, this imprinting seems to involve specifically MC4-R in the VMH, as a significant correlation was identified between receptor density here at 8 weeks and plasma leptin levels at 1 week after presentation of palatable food (Figure 3). We therefore suggested that those individuals able to elevate plasma leptin - which presumably stimulates a subset of POMC neurons projecting to impact on the MC4-R in the VMH - are able to increase tonic activation of these receptors and thus protect themselves against over-eating and developing obesity.

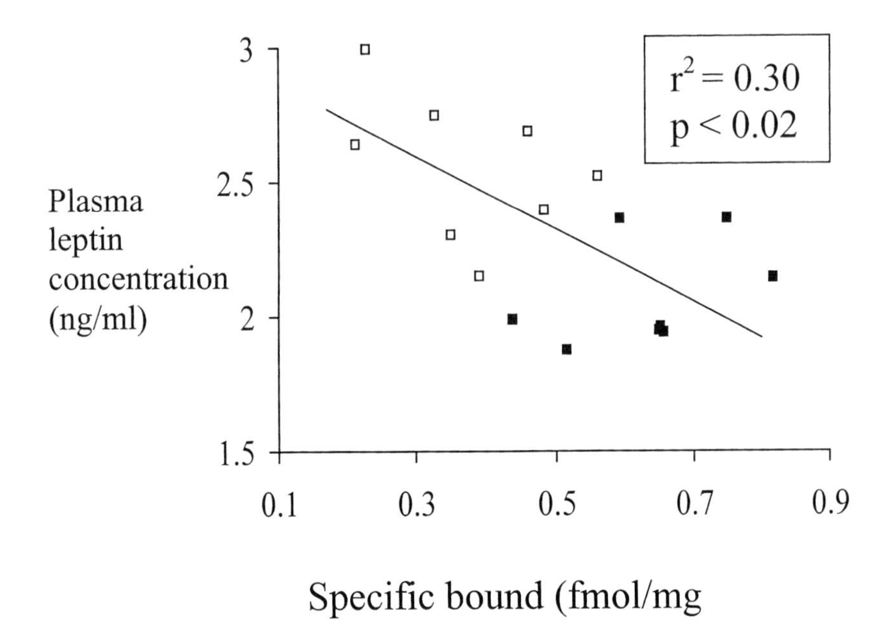

Figure 3. Correlation between plasma leptin concentration measured after 1 week of energy-dense diet feeding and the density of MC4-R in the VMH of obesity-prone and obesity-resistant rats after 8 weeks. Key: obesity-resistant (χ), obesity-prone (γ). (Adapted from Harrold *et al.*, 2000a, by permission of the author)

Various genetic, physiological and pharmacological studies have therefore defined a role for MC4-R in the regulation of energy homeostasis and determination of susceptibility to obesity. The physiological involvement of MC3-R controversial. The MC3-R knockout mouse model demonstrates increased feeding efficiency i.e. the amount of weight gained and fat deposited per gram of food eaten, but in the presence of *decreased* food intake. Consequently, animals have increased body fat mass, but reduced lean mass such that there is no overall difference in body weight (Chen *et al.*, 2000). As yet, the pathophysiological mechanism underlying this phenotype are unknown as MC3-R knockout mice do not show changes in metabolic rats or body temperature. However, they are relatively inactive and this could, in part, explain the obesity (Chen *et al.*, 2000). As these changes occur in the presence of decreased food intake these animals should be unusually susceptible to DIO, but this hypothesis has not yet been tested

The melanocortin receptors are excellent candidates to be thrifty genes: unlike leptin, allelic variations should promote weight gain and fat storage without impairing fertility and thus preventing positive selection of the mutation. In man it has been confirmed that certain mutations in MC4-R predispose to obesity. Recessive inheritance of such mutations are also associated with obesity. In fact, various MC4-R mutations have been identified at relatively high prevalance in certain human populations and are proposed to be responsible for perhaps 4% of severe obesity in these groups (Farooqi *et al.*, 2000). The contribution of the MC3-R to determining susceptibility to human obesity is enigmatic at present; despite the promising phenotype of the MC3-R knockout mouse, initial human studies indicate that there is no linkage between MC3-R mutations and obesity (Li *et al.*, 2000). Further investigation is obviously required to confirm or refute this.

Reward Signals

There is considerable evidence to support the idea that neuronal circuits innervating the hypothalamus are critical to the control of energy homeostasis, but regulatory systems in other brain areas are also likely to play a role. Recent work suggests that components of the latter systems may also influence susceptibility to obesity, notably by influencing palatability or the 'hedonic' aspects of food. The current theory is that the amount consumed of a given food is governed by competition between the reward value generated by eating it *vs* the hypothalamic generated satiety signals. An unbalance of these neurochemical signals such that the reward stimuli dominate could theoretically provide a drive to over-eat and cause obesity in dietary-susceptible rats. Therefore, neurons activated by ingestion of palatable energy-rich foods are also prime targets for the identification of potential thrifty genes.

Forebrain regions, especially the nucleus accumbens, are known to be important in motivational processes and reinforcement behaviour (Wize and Bozarth, 1987). Several neurotransmitters may act in this region to modulate feeding, including the dopamine, opioid and nociceptin systems which densely innervate the forebrain. Of these, the opioid system currently seems most relevant in determining susceptibility to DIO.

It has previously been suggested that disturbed opioid activity may be linked to disorders characterised by excessive eating and impulsive behaviour (Kelley *et al.*, 1996). Acute administration of opioid receptor agonists and antagonists stimulates and inhibits food intake respectively (Morely *et al.*, 1983) and some evidence suggests that the peptides specifically influence fat consumption and thus may modulate palatability (Marks-Kaufman, 1982). Furthermore, opioid peptides and their receptors may partly determine interstrain variability in susceptibility to obesity. β-casomorphine$_{(1-7)}$, an opioid like peptide, increased consumption of high-fat food but not low-fat diet in Osborne-Mendel rats susceptible to dietary obesity, but had no influence on food intake in the S5B/P1 strain which are resistant to dietary obesity (Lin *et al.*, 1996).

However, we undoubtedly remain naïve regarding the mechanisms underlying the hedonic aspects of ingestive behaviour. Further investigation of the opioids and other neuropeptides acting within the nucleus accumbens and surrounding regions may improve understanding of their role in appetite control and ultimately define genetic variabilities in the system responsible for influencing susceptibility to obesity.

Adaptive Thermogenesis

Body weight is maintained by a fine balance between energy intake and energy expenditure. Various metabolic adaptations exist that enable an individual to dissipate energy as heat following excessive over-eating, in an attempt to limit weight gain. One important adaptive mechanism is an increase in metabolic rate. In rodents this has been largely attributed to enhanced energy expenditure by brown adipose tissue (BAT), the principal thermogenic tissue, and in particular to mitochondrial uncoupling proteins (UCPs) present within it.

UCP1 is expressed solely in BAT and functions to short circuit protons (generated by free fatty acid oxidation) back into the mitochondrial matrix, resulting in heat production rather than ATP synthesis (Ricquier *et al.*, 1991). BAT is clearly important in regulating energy expenditure as its ablation (by transgenic expression of diptheria toxin) leads to obesity (Lowell *et al.*, 1993). Interestingly, however, UPC1 knockout mice to not become obese, but are cold intolerant, suggesting that UCP1 induction in BAT is important for generating heat primarily during periods of cold exposure (Enerback *et al.*, 1992). Up-regulation of other UCPs in BAT may take over from UCP1 under conditions of normal ambient temperature. Increased activity of UCP1 has been proposed to underpin the increased thermogenesis of obesity (Rothwell and Stock, 1986) and up-regulation of UCP1 has been reported in young animals fed an energy-rich diet (Rothwell and Stock, 1983). However, this phenomenom is not observed in older rats despite their elevated body temperature (Harrold *et al.*, 2000b). The low abundance of BAT and of UCP-1 in adult humans also implies that this is a factor of limited potential in influencing susceptibility to DIO.

Recent studies have reported two related proteins, UCP2 and UCP3, that may be more relevant to body weight regulation. UCP3 is expressed predominantly in muscle and BAT (Boss *et al.*, 1997) whereas UCP2 is expressed abundantly in many tissues (Fleury *et al.*, 1997). Differences in UCP2 regulation may contribute to variability between strains of mice in susceptibility to DIO, as the dietary-resistant mouse strain A/J can up-regulate UCP2 in white adipose tissue more markedly that the dietary-susceptible C57BL/6 strain (Surwit *et al.*, 1998). However, this role for UCP2 is also controversial, as it has been shown that no variations in either UCP2 or UCP3 regulation occur between dietary-susceptible and dietary-resistant animals within the same rat strain, upon exposure to a palatable diet (Figure 4; Harrold *et al.*, 2000b).

Research into the contribution of UCPs to determining susceptibility to DIO is still in its early years. UCP2 (Arsenijevic *et al.*, 2000) and

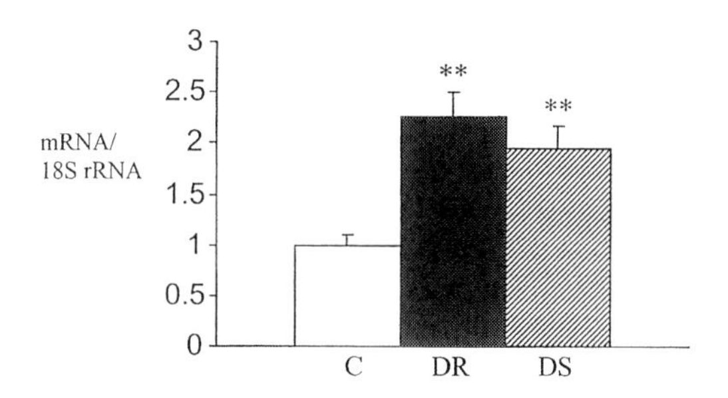

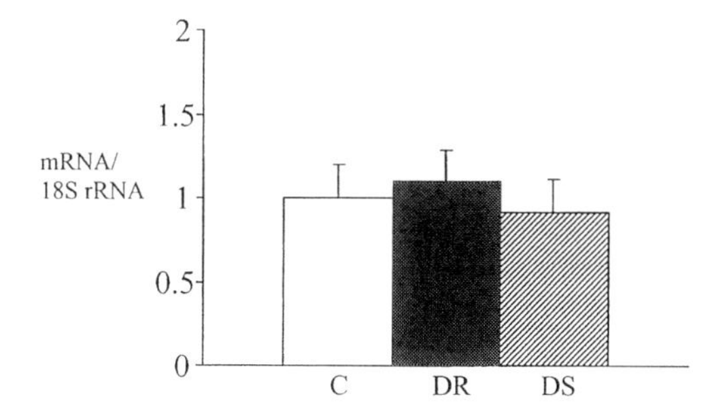

Figure 4. UCP3 mRNA levels in gastrocnemius muscle and UCP2 mRNA levels in white adipose tissue (WAT) do not differ between obesity-resistant (DR) and obesity-prone (DS) rats fed an energy dense diet for 8 weeks. (Adapted from Harrold *et al.*, 2000b, by permission of the author)

UCP3 (Vidal-Puig *et al.,* 2000) knockout mice have recently been generated. Neither model is obese and both demonstrate normal responses to cold exposure and high-fat diet. However, the contribution of alternative corrective mechanisms to these phenotypes have not yet been excluded. Further research will therefore be required to confirm their function and possible genetic contribution to thermogenesis and metabolic rate. Crucially, their relevance to human obesity remains to be determined. Population studies have pointed to some polymorphisms in the UCP-2 and UCP-3 genes being associated with disturbed weight regulation in some cases but not in others, and no convincing candidates have yet emerged. Moreover, to date genetic studies have not established any loss of function mutations in obese individuals (Walder *et al.*, 1998).

NEURONAL PLASTICITY

The notion of 'metabolic imprinting' or programming of neural circuits is relatively novel. This theory proposes that the formation of new neuronal circuits or the alteration of existing ones can occur in individuals genetically predisposed to obesity upon exposure to energy-dense food and/or as a consequence of maternal influences. There are other precedents in neurobiology. The formation of long-term memory implies that repeated use of new synaptic connections strengthens them and eventually renders them permanent by gene induction (Bailey and Chen, 1983). Additionally, this process is modulated by neurotransmitters and hormones, including many that are known to be altered in dietary-susceptible animals. According to this model, the activity of new synaptic connections in the circuits controlling energy homeostasis would effectively defend a higher body weight because the body weight set-point would drift up to a higher level during the chronic over consumption of energy. Indications that this process does occur include reports of reduced numbers of axonal synapses on ARC cell bodies of adult DIO rats (Levin, 2000), reduced numbers of forebrain α_2-adrenoreceptors (Wilmot *et al.*, 1988) and reduced neuronal activity in the VMH (Levin and Sullivan, 1989b). However, there is no firm evidence available as yet to demonstrate that the formation of new and permanent neural connections plays a role in the development or maintenance of obesity, as distinct from the impressive but essentially transient alterations in the expression of neurotransmitters and their receptors mentioned above.

CONCLUSION

The past few years have seen substantial progress in our understanding of the molecular mechanisms of body weight regulation. Significant advances include the identification of mutations in components of the leptin and melanocortin pathways in obese humans. However, rodents continue to be the central focus of study where ongoing work will help to clarify the mechanisms that control food intake and energy homeostasis. In turn, such information should provide a more detailed understanding of the pathogenesis of human obesity and ultimately lead to the identification of therapeutic strategies to cut the cost of this disease on human health.

REFERENCES

Arsenijevi D, Onuma H, Pecqueur C, Raimbault S, Manning BS, Miroux B, Couplan E, Alves-Guerra M-C, Goubern M, Surwit R, Bouillard F, Richard D, Collins S, Ricquier D. Disruption of the uncoupling protein-2 gene in mice reveals a role in immunity and reactive oxygen species production. *Nat Genet* 2000; 26: 435-439.

Bailey CH, Chen M. Morphological basis of long term habituation and sensitization in aplysia. *Science* 1983; 220: 91-93.

Balasubramaniam AA. Neuropeptide Y family of hormones: receptor subtypes and antagonists. *Peptides* 1997; 18: 445-457.

Berenson GS, Bao W, Srinivasan SR. Abnormal characteristics of young offspring of parents with non-insulin-dependent diabetes mellitus. The Bogalusa Heart Study. *Am J Epidemiol* 1997; 144: 962-967.

Bjorbaek C, Elmquist JK, Frantz JD, Shoelson SE, Flier JS. Identification of SOCS-3 as a potential mediator of central leptin resistance. *Mol Cell* 1998; 1(4): 619-625.

Boss O, Samec S, Paoloni-Giacobino A, Rossier C, Dulloo A, Seydoux J, Muzzin P, Giacobino J-P. Uncoupling protein-3: a new member of the mitochondrial carrier family with tissue specific expression. *FEBS Lett* 1997; 408: 39-42.

Bouchard C, Tremblay A. Genetic influences on the response of body fat and fat distribution to positive and negative energy balances in human identical twins. *J Nutr* 1997; 127: 943S-947S.

Bray G.A. Obesity, a disorder of nutrient partitioning: the MONA LISA hypothesis. *J Nutr* 1991a; 121: 1146-1162.

Bray GA. Reciprocal relation between the sympathetic nervous system and food intake. *Brain Res Bull* 1991b; 27: 517-520.

Bray GA York DA. Hypothalamic and genetic obesity in experimental animals: an autonomic and endocrine hypothesis. *Physiol Rev* 1979; 59(3): 719-809.

Broberger C, Johansen J, Johansson C, Schalling M, Hokfelt T. The neuropeptide Y/agouti gene-related protein (AGRP) brain circuitry in normal, anorectic and monosodium glutamate-treated mice. *Proc Natl Acad Sci USA.* 1998; 95: 15043-15048.

Burguera B, Conce ME, Curran GL, Jensen MD, Lloyd RV, Cleary MP, Poduslo JF. Obesity is associated with decreased leptin transport across the blood-brain barrier in rats. *Diabetes* 2000; 49(7): 1219-1223.

Campfield LA, Smith FJ. Modulation of insulin secretion by the autonomic nervous system. *Brain Res Bull* 1980; 5: 103-107.

Caro JF, Kolaczynski JW, Nyce MR, Ohannesian J.P, Opentanova I, Goldman WH, Lynn RB, Zhang PL, Sinha MK, Considine RV. Decreased cerebrospinal-fluid/serum leptin ratio in obesity: a possible mechanism for leptin resistance. *Lancet* 1996; 348: 159-161

Chang S, Graham B, Yakabu F, Lin D, Peters JC, Hill JO. Metabolic differences between obesity-prone and obesity-resistant rats. *Am J Physiol* 1990; 259: R1103-R1110.

Chen H, Charlat O, Tartaglia LA, Woolf EA, Weng X, Ellis SJ, Lakey ND, Culpepper J, Moore KJ, Breitbart RE, Duyk GM, Tepper RI, Morgenstern JP. Evidence that the diabetes gene encodes the leptin receptor; identification of a mutation in the leptin receptor gene in *db/db* mice. *Cell* 1996; 84: 491-495.

Chen AS, Marsh DJ, Trumbauer ME, Frazier E.G, Guan X-M, Yu H, Rosenblum CI, Vongs A, Feng Y, Cao L, Metzger JM, Strack AM, Camacho RE, Mellin TN, Nunes CN, Min WM, Fisher J, Gopal-Truter S, MacIntyre DE, Chen HY, Van der Ploeg LHT. Inactivation of the mouse melanocortin-3 receptor results in increased fat mass and reduced lean body mass. *Nat Genet* 2000; 26: 97-102.

Comuzzie AG Allison DB. The search for the human obesity genes. *Science* 1998; 280: 1374-1377.

Depocas F, Behrens WA, Foster DO. Noradrenaline-induced calorigenesis in warm- and cold-acclimated rats: the inter-relationship of dose of noradrenaline, its concentration in the arterial plasma and calorigenic response. *Can J Physiol Pharmacol* 1978; 56: 168-174.

Enerback S, Jacobsson A, Simpson EM, Guerra C, Yamashita H, Harper ME, Kozak LP. Mice lacking mitochondrial uncoupling protein are cold-sensitive but not obese. *Nature* 1997; 387: 90-94.

Fan W, Boston BA, Kesterson RA, Hruby VJ, Cone RD. Role of melanocortinergic neurons in feeding and the *agouti* obesity syndrome. *Nature* 1997; 385: 165-168.

Farooqi IS, Jebb SA, Langmack G, Lawrence E, Cheetham CH, Prentice AM, Hughes IA, McCamish MA, O'Rahilly S. Effects of recombinant leptin therapy in a child with congenital leptin deficiency. *N Engl J Med* 1999; 341(12): 879-884.

Farooqi IS, Yeo GS, Keogh JM, Aminian S, Jebb SA, Butler G, Cheetham T, O'Rahilly S. Dominant and recessive inheritance of morbid obesity associated with melanocortin-4 receptor deficiency. *J Clin Invest* 2000; 106(2): 271-279.

Faust IM, Johnson PR, Hirsch J. Long-term effects of early nutritional experience on the development of obesity in the rat. *J Nutr* 1980; 110(10): 2027-2034.

Fleury C, Neverova M, Collins S, Raimbault S, Champigny O, Levi-Meyrueis C, Bouillard F, Seldin MF, Surwit RS, Ricquier D, Warden CH. Uncoupling protein-2: a novel gene linked to obesity and hyperinsulinaemia. *Nat Genet* 1997; 15: 269-272.

Folli F, Ghidella S, Bonfanti L, Kahn CR, Merighi A. The early intracellular signaling pathway for the insulin/insulin-like growth factor receptor family in the mammalian central nervous system. *Mol Neurobiol* 1996; 13: 155-183.

Foster DO, Frydman ML. Nonshivering thermogenesis in the rat. II. Measurement of blood flow with microspheres point to brown adipose tissue as the dominant site of calorigenesis induced by noradrenaline. *Can J Physiol* 1978; 56: 110-122.

Frederich RC, Arner P, Nordfors L, Schalling M. Leptin levels reflect body lipid content in mice: evidence for diet-induced resistance to leptin action. *Nat Med* 1995; 1: 1311-1314.

Friedman JM, Halaas JL. Leptin and the regulation of body weight in mammals. *Nature* 1998; 395: 763-770.

Gautier JF, Chen K, Bandy D. Increased cortical representation of hunger and satiation in obese men. *Diabetes* 1999; 48(Suppl. 1): A312.

Godbole V, York DA, Bloxham DP. Developmental changes in the fatty (fa/fa) rat: evidence for defective thermogenesis preceding the hyperlipogenesis and hyperinsulinaemia. *Diabetologia* 1978; 15: 41-44.

Hamilton BS, Paglia D, Kwan AYM, Deitel M. Increased *obese* mRNA expression in omental fat cells from massively obese humans. *Nat Med* 1995; 1: 953-956.

Harrold JA, Williams G, Widdowson PS. Early leptin response to a palatable diet predicts dietary obesity in rats: key role of melanocortin-4 receptors in the ventromedial hypothalamic nucleus. *J Neurochem* 2000a; 74: 1224-1228.

Harrold JA, Widdowson PS, Clapham JC, Williams G. Individual severity of dietary obesity in unselected wistar rats: relationship with hyperhagia. *Am J Physiol* 2000b; 279: E340-E347.

Heidel E, Plagemann A, Davidowa H. Increased response to NPY of hypothalamic VMN neurons in postnatally overfed juvenile rats. *Neuroreport* 1999; 10(9): 1827-1831.

Hill JO, Fried SK, DiGirolamo M. Effects of a high-fat diet on energy intake and expenditure in rats. *Life Sci* 1983; 33(2): 141-149.

Huszar D, Lynch CA, Fairchild-Huntress V, Dunmore JH, Fang Q, Berkemeier LR, Gu W, Kesterson RA, Boston BA, Cone R.D, Smith FJ, Campfield LA, Burn P, Lee F. Targeted disruption of the melanocortin-4 receptor results in obesity in mice. *Cell* 1997; 88: 131-141.

Jones AP, Assimon SA, Friedman MI. The effect of diet on food intake and adiposity in rats made obese by gestational undernutrition. *Physiol Behav* 1986; 37: 381-386.

Kelley A.E, Lang CG, Gauthier AM. Induction of oral stereotypy following amphetamine microinjection into a discrete subregion of the striatum. *Psychopharmacology* 1988; 95: 556-559.

Leibowitz SF, Alexander JT. Analysis of neuropeptide Y-induced feeding: dissociation of Y1 and Y2 receptor effects on natural meal patterns. *Peptides* 1991; 12: 1251-1260.

Levin BE. Intracarotid glucose-induced norepinephrine response and the development and maintenance of diet-induced obesity. *Int J Obes* 1992; 16: 451-457.

Levin BE. Sympathetic activity, age, sucrose preference and diet-induced obesity. *Obese Res* 1993; 1: 281-287.

Levin BE. Reduced norepinephrine turnover in organs and brains of obesity-prone rats. *Am J Physiol* 1995; 268: R389-R394.

Levin BE, Dunn-Meynell A. Defense of body weight against chronic caloric restriction in obesity-prone and obesity-resistant rats. *Am J Physiol* 2000; 278: R231-R238.

Levin BE, Govek E. Gestational obesity accentuates obesity in obesity-prone progeny. *Am J Physiol* 1998; 275: R1375-R1379.

Levin BE, Planas B. Defective glucoregulation of brain α_2-adrenoreceptors in obesity-prone rats. *Am J Physiol* 1993; 264: R305-R311.

Levin BE, Sullivan AC. Glucose-induced norepinephrine levels and obesity resistance. *Am J Physiol* 1987; 253: R475-R481.

Levin BE, Sullivan AC. Glucose-induced sympathetic activation in obesity-prone and obesity-resistant rats. *Int J Obes* 1989a ; 13: 235-246.

Levin BE, Sullivan AC. Differences in saccharin-induced cerebral glucose utilization between obesity-prone and resistant rats. *Brain Res* 1989b; 488: 221-232.

Levin BE, Comai K, Sullivan AC. Metabolic sympatho-adrenal abnormalities in the obese Zucker rat: effect of chronic phenoxybenzamine treatment. *Pharmacol Biochem Behav* 1981; 14: 517-525.

Levin BE, Dunn-Meyell AA, Balkan B, Keesey RE. Selective breeding for diet-induced obesity and resistance in Sprague-Dawley rats. *Am J Physiol* 1997; 273: R725-R730.

Levin BE, Israel P, Lattemann DP. Insulin selectively downregulates alpha-2-adrenoreceptors in the arcuate and dorsomedial nucleus. *Brain Res Bull* 1998; 45(2): 179-181.

Levin BE, Triscari J, Hogan S, Sullivan A. Resistance to diet-induced obesity: food intake, pancreatic sympathetic tone and insulin. *Am J Physiol* 1987; 252: R471-R478.

Levin BE, Triscari J, Sullivan AC. Relationship between sympathetic activity and diet-induced obesity in two rat strains. *Am J Physiol* 1983; 245: R367-R371.

Levin BE, Triscari J, Sullivan A. Metabolic features of diet-induced obesity without hyperphagia in young rats. *Am J Physiol* 1986; 251: R433-R440.

Li WD, Joo EJ, Furlong EB, Galvin M, Abel K, Bell CJ, Price RA. Melanocortin 3 receptor (MC3-R) gene variants in extremely obese women. *Int J Obes Relat Metab Disord* 2000; 2: 206-210.

Lin L, York DA, Bray GA. Comparison of Osborne-Mendel and S5B/PL strains of rats: Central effects of galanin, NPY, β-Casomorphin and CRH on intake of high-fat and low-fat diets. *Obes Res* 1996; 4(2): 117-124.

Lowell BB, Susulic S-V, Hamann A, Lawitts JA, Himms-Hagen J, Boyer BB, Kozak LP, Flier JS. Development of obesity in transgenic mice after genetic ablation of brown adipose tissue. *Nature* 1993; 366: 740-742.

Lu D, Willard D, Patel IR, Kadwell S, Overton T, Kost M, Luther W, Chen W, Woychik RP, Wilkinson WO, Cone RD. Agouti protein is an antagonist of the melanocyte-stimulating hormone receptor. *Nature* 1994; 371: 799-802.

Madiehe AM, Schaffhauser AO, Braymer DH, Bray GA, York DA. Differential expression of leptin receptor in high- and low-fat-fed Osborne-Mendel and S5B/P1 rats. *Obes Res* 2000; 8(6): 467-474.

Maffei M, Fei H, Lee G-H, Dani C, Leroy P, Zhang Y, Proenca R, Negrel R, Ailhaud G, Friedman JM. Increased expression in adipocytes of ob RNA in mice with lesions of the hypothalamus and with mutations in the *db* locus. *Nat Genet* 1995; 92: 6957-6960.

Marks-Kaufman R. Increased fat consumption induced by morphine administration in rats. *Pharmacol Biochem Behav* 1982; 16: 949-955.

Mokdad AH, Serdula MK, Dietz WH, Bowman BA, Marks JS, Koplan JP. The spread of the obesity epidemic in the United States, 1991-1998. *JAMA* 1999; 282: 1519-1522.

Montague CT, Farooqi IS, Whitehead JP, Soos MA, Rau H, Wareham NJ, Sewter C, Digby JE, Mohammed SN, Hurst JA, Cheetham CH, Earley AR, Barnett AH, Prins JB, O'Rahilly S. *Nature* 1997; 387: 903-908.

Morley JE, Levine AS, Yim GKW, Lowy MT. Opioid modulation of appetite. *Neurosci Biobehav Rev* 1983; 7: 281-305.

Mountjoy KG, Mortrud MT, Low MJ, Simerly RB, Cone RD. Localization of the melanocortin-4 receptor (MC4-R) in neuroendocrine and autonomic control circuits in the brain. *Mol Endocrinol* 1994; 8: 1298-1308.

Nataf V, Monier S. Effect of insulin and insulin-like growth factor I on the expression of the catecholaminergic phenotype by neural crest cells. *Brain Res Dev Brain Res* 1992; 1: 59-66.

Neel JV. Diabetes mellitus: a thrifty genotype rendered detrimental by 'progress'? *Am J Hum Genet* 1962; 14: 353-362.

Ollmann MM, Wilson BD, Yang Y-K, Kerns JA, Chen Y, Gantz I, Barsh GS. Antagonism of central melanocortin receptors in vitro and in vivo by agouti-related protein. *Science* 1997; 278: 135-138.

O'Shea D, Morgan DG, Meeran K, Edwards CM, Turton MD, Choi SJ, Heath MM, Gunn I, Taylor GM, Howard JK, Bloom CI, Small CJ, Haddo O, Ma JJ, Callinan W, Smith DM, Ghatei MA, Bloom SR. Neuropeptide Y induced feeding in the rat is mediated by a novel receptor. *Endocrinol* 1997; 138: 196-202.

Perseghin GS, Ghosh S, Gerow K, Shulman GI. Metabolic defects in lean nondiabetic offspring of NIDDM parents: a cross-sectional study. *Diabetes* 1997; 46: 1001-1009.

Plagemann A, Heidrich I, Gotz F, Rohde W, Dorner G. Lifelong enhanced diabetes susceptibility and obesity after temporary intrahypothalamic hyperinsulinism during brain organization. *Exp Clin Endocrinol* 1992; 99: 91-95.

Plagemann A, Harder T, Rake A, Melchior K, Rohde W, Dorner G. Increased number of galanin-neurons in the paraventricular hypothalamic nucleus of neonatally overfed weanling rats. *Brain Res* 1999a; 818(1): 160-163.

Plagemann A, Harder T, Rake A, Waas T, Melchior K, Ziska T, Rohde W, Dorner G. Observations on the orexigenic hypothalamic neuropeptide Y-system in neonatally overfed weanling rats. *J Neuroendocrinol* 1999b; 11(7): 541-546.

Plagemann A, Harder T, Melchior K, Rake A, Dorner G. Elevation of hypothalamic neuropeptide Y-neurons in adult offspring of diabetic mother rats. *Neuroreport* 1999c; 10(15): 3211-3216.

Plagemann A, Harder T, Rake A, Janert U, Melchior K, Rohde W, Dorner G. Morphological alterations of hypothalamic nuclei due to intrahypothalamic hyperinsulinism in newborn rats. *Int J Dev Neurosci* 1999d; 17(1): 37-44.

Puro DG, Agardh E. Insulin-mediated regulation of neuronal maturation. *Science* 1984; 225: 1170:1172.

Ravussin E, Pratley RE, Maffei M, Wang H, Friedman JM, Bennett PH, Bogardus C. Relatively low plasma leptin concentrations precede weight gain in Pima Indians. *Mat Med* 1997; 3: 238-240.

Reid JM, Fullmer SD, Pettigrew KD, Burch TA, Bennett PH, Miller M, Whedon GD. Nutrient intake of Pima Indian women: relationship to diabetes mellitus and gallbaldder disease. *Am J Clin Nutr* 1971; 24(10): 1281-1289.

Ricquier D, Casteilla L, Bouillard F. Molecular studies of the uncoupling protein. *FASEB J* 1991; 5: 2237-2242.

Rossi M, Kim MS, Morgan DG, Small CJ, Edwards CM, Sunter D, Abunsnana S, Goldstone AP, Russell SH, Stanley SA, Smith DM, Yagaloff K, Ghatei MA, Bloom S.R. A C-terminal fragment of agouti-related protein increases feeding and antagonizes the effect of alpha-melanocyte stimulating hormone in vivo. *Endocrinol* 1998; 139: 4428-4431.

Rothwell NJ, Stock MJ. Effects of age on diet-induced thermogenesis and brown adipose tissue metabolism in the rat. *Int J Obes* 1983; 7: 583-589.

Rothwell NJ, Stock MJ. Brown adipose tissue and diet-induced thermogenesis. In: *Brown Adipose Tissue*. London: Edward Arnold, p. 269-298. 1986.

Sahu A, Kalra PS, Kalra SP. Food deprivation and ingestion induced reciprocal changes in neuropeptide Y-induced feeding in rats. *Peptides* 1988; 9: 83-86.

Sanacora G, Kershaw M, Finkelstein JA, White JD. Increased hypothalamic content of preproneuropeptide Y messenger ribonucleic acid in genetically obese Zucker rats and its regulation by food deprivation. *Endocrinol* 1996; 127: 730-737.

Schaffhauser AO, Stricker-Krongrad A, Brunner L, Cumin F, Gerald C, Whitebread S, Criscione L, Hofbauer KG. Inhibition of food intake by neuropeptide Y Y5 receptor antisense oligodeoxynucleotides. *Diabetes* 1997; 46: 1792-1798.

Schwartz MW, Woods SC, Porte D, Seeley RJ, Baskin DG. Central nervous system control of food intake. *Nature* 2000; 404: 661-671.

Shutter JR, Graham M, Kinsey AC, Scully S, Luthy R, Stark KL. Hypothalamic expression of ART, a novel gene related to agouti, is up-regulated in *obese* and *diabetic* mutant mice. *Genes Dev* 1997; 11: 593-602.

Silverman BL, Metzger BE, Cho NH, Loeb CA. Impaired glucose tolerance in adolescent offspring of diabetic mothers: relationship to fetal hyperinsulinism. *Diabetes Care* 1995; 18: 611-617.

Spraul M, Ravussin E, Fontvieille AM, Rising R, Larson DE, Anderson EA. Reduced sympathetic nervous activity. A potential mechanism predisposing to body weight gain. *J Clin Invest* 1993; 92(4): 1730-1735.

Stanley BG, Kyrhouli SE, Lampert S, Leibowitz SF. Neuropeptide Y chronically injected into the hypothalamus: a powerful neurochemical inducer of hyperphagia and obesity. *Peptides* 1993; 7: 1189-1192.

Surwit RS, Petro AE, Parekh P, Collins S. Low plasma leptin in response to dietary fat in diabetes- and obesity-prone mice. *Diabetes* 1997; 46: 1516-1520.

Surwit RS, Wang S, Petro AE, Sanchis D, Raimbault S, Ricquier D, Collins S. Diet-induced changes in uncoupling proteins in obesity-prone and obesity-resistant strains of mice. *Proc Natl Acad Sci USA*. 1998; 95: 4061-4065.

Surwit RS, Edwards CL, Murphy S, Petro AE. Transient effects of long-term leptin supplementation in the prevention of diet-induced obesity in mice. *Diabetes* 2000; 49(7): 1203-1208.

Tanaka M, Sawada M, Yoshida S, Hanaoka F, Marunouchi T. Insulin prevents apoptosis of external granular layer neurons in rat cerebellar slice cultures. *Neurosci Lett* 1995; 199: 37-40.

Vidal-Puig AJ, Grujic D, Zhang C-Y, Hagen T, Boss O, Ido Y, Szczepanik A, Wade J, Mootha V, Cortright R, Muoio DM, Lowell BB. Energy metabolism in uncoupling protein 3 gene knockout mice. *J Biol Chem* 2000; 275(21): 16258-16266.

Walder K, Norman RA, Hanson RL, Schrauwen P, Neverova M, Jenkinson CP, Easlick J, Warden CH, Pecqueur C, Raimbault S, Ricquier D, Silver MHK, Shuldiner AR, Solanes G, Lowell BB, Chung WK, Leibel RL, Prately R, Ravussin E. Association between uncoupling protein polymorphisms (UCP2-UCP3) and energy metabolism/obesity in Pima Indians. *Hum Mol Genet* 1998; 9: 1431-1435.

Wang J, Liu R, Hawkins M, Barzilia N, Rossetti L. A nutrient-sensing pathway regulates leptin gene expression in muscle and fat. *Nature* 1998; 393: 684-688.

Widdowson PS, Upton R, Henderson L, Buckingham R, Wilson S, Williams G. Reciprocal regional changes in brain NPY receptor density during dietary restriction and dietary-induced obesity in the rat. *Brain Res* 1997; 774: 1-10.

Wilding JP, Gilbey SG, Bailey CJ, Batt RA, Williams G, Ghatei MA, Bloom SR. Increased neuropeptide-Y messenger ribonucleic acid (mRNA) and decreased neurotensin mRNA in the hypothalamus of the obese *ob/ob* mouse. *Endocrinol* 1993; 132: 1939-1944.

Wilmot CA, Sullivan AC, Levin BE. Effects of diet and obesity on brain A_1- and A_2-noradrenergic receptors in the rat. *Brain Res* 1988; 453: 157-166.

Wise RA, Bozarth MA. A psychomotor stimulant theory of addiction. *Psychol Rev* 1987; 94: 469-492.

Wright PH, Malaisse IJ. Effects of epinephrine, stress and exercise on insulin secretion by the rat. *Am J Physiol* 1968; 214: 1031-1034.

Yamaguchi T, Keino K, Fukuda J. The effect of insulin and insulin-like growth factor-1 on the expression of calretinin and calbindin D-28k in rat embryonic neurons in culture. *Neuro-chem Int* 1995; 26: 255-262.

Zucker LM, Antoniades HN. Insulin and obesity in the Zucker genetically obese rat 'fatty'. *Endocrinol* 1972; 90: 1320-1330.

PART 4

Genetic susceptibility to leanness in animals

Chapter 9

MUSCLE ENHANCED TRAITS IN CATTLE AND SHEEP

Noelle E. Cockett and Christopher A. Bidwell
Utah State University, Department of Animal, Dairy and Veterinary Sciences, Logan, UT 84322-4700 USA; Purdue University, Department of Animal Sciences, West Lafayette, IN 47907-1026 USA

ABSTRACT The development of skeletal muscle involves proliferation of muscle precursor cells (myoblasts) and subsequent fusion into myotubes or muscle fibers. Therefore, skeletal muscle growth requires both an increase in cell numbers and an increase in muscle fiber size. Natural mutations exist in both cattle and sheep that affect either proliferation of myoblasts or muscle fiber size. These mutations have been retained in certain breeds due to selection. Inactivation of the *myostatin* locus in mice and in double muscled cattle has identified an important new negative regulator in skeletal muscle growth. Myostatin has been shown to be important for fetal muscle development and in maintenance of muscle mass in adults. Two loci on sheep chromosome 18 (*callipyge* and *ribeye muscle*) are linked to enhanced postnatal skeletal muscle growth in the pelvic limbs and loin. However, the actual mechanism responsible for muscle enhancement in these traits is not yet known. These mutations in livestock will provide new insight into the control of growth and body composition.

INTRODUCTION

Growth in a population of cells is an interaction between extracellular signals and intracellular programs that increase cell numbers (hyperplasia) or increase cell size (hypertrophy). In skeletal muscle, hyperplasia is due to proliferation of mononucleated muscle progenitor cells called myoblasts. The myoblasts undergo terminal differentiation, exit the cell cycle, and fuse into new multinucleated myotubes or an existing muscle fiber. The multinucleated cell undergoes hypertrophy by the synthesis and accumulation of greater amounts of myofibrillar

J.B. Owen et al. (eds.), Animal Models – Disorders of Eating Behaviour and Body Composition, 159–172.

proteins and addition of new sarcomeres, causing the myofibers to grow in length and diameter. The bovine *muscle hypertrophy* locus (*mh*), commonly known as double muscling, causes a near doubling of muscle fiber numbers during fetal development, presumably due to myoblast hyperplasia. Calves are born with larger muscles and have associated birthing difficulties. In contrast, the ovine callipyge mutation (*CLPG*) produces a trait where the number of muscle fibers remains constant but the fibers grow significantly larger after birth. Therefore, callipyge animals have enhanced muscle definition primarily due to hypertrophy.

Myogenesis, the formation of muscle fibers, occurs through a carefully orchestrated process of proliferation and differentiation. This process is facilitated by muscle regulatory factors (MRFs), which belong to a family of basic helix-loop-helix transcription factors that initiate myogenesis and regulate the transcription of muscle specific genes (reviewed by Weintraub, 1993). In the first step of proliferation, embryonic mesoderm cells from the somites are recruited to the muscle lineage because of extracellular signaling by the secreted proteins sonic hedgehog and Wnt (Cossu and Borello,1999). These factors induce the expression of pax-3 which then induces the expression of the MFRs, myf-5 and MyoD (Tajbakhsh et al., 1997; Maroto et al., 1997). These cells continue to proliferate, forming populations of myoblasts which then express myogenin, another MRF. At this point the cells are called myocytes and are committed to terminal muscle differentiation. Activation of the cyclin-dependent kinase inhibitors p21 and p27 in the myocytes (Bennet and Tonks,1997; Walsh and Perlman,1997) results in a permanent exit from the cell cycle followed by fusion into myotubes. Chromatin remodeling within the myotubes then occurs, which facilitates gene transcription of several muscle specific genes such as the myosin and actin contractile proteins and a variety of cell surface proteins such as integrins, fibronectins and neural cell adhesion molecules (Walsh and Perlman,1997). A final stage of differentiation, referred to as maturation, occurs as the primary myofibers and secondary myofibers begin to express different isoforms of myosin heavy chain genes. Each muscle or group of muscles has a characteristic distribution of Type I and Type II fibers, although there is a great deal of plasticity in fiber type depending on neural inputs, hormonal stimulation or exercise demands (reviewed by Buonanno and Rosenthal,1996).

Three distinct populations of myoblasts, (embryonic, fetal and adult) exist at different times during embryonic and postnatal development of skeletal muscle (Stockdale,1992). Embryonic myoblasts undergo proliferation, migration, and fusion to form primary myotubes that will ultimately determine the muscle pattern. The second population, called fetal myoblasts, fuse to form secondary muscle fibers around the scaffold of primary muscle fibers. In Drosophila, a transmembrane cell adhesion protein, called dumbfounded, is responsible for organizing myotube fusion. Expression of the *dumbfounded* gene by a subset of "founder" myoblasts is an attractant for other fusion capable myoblasts and is necessary for their aggregation prior to fusion into myotubes (Ruiz-Gomez et al., 2000). Adult myoblasts, which are also referred to as satellite cells, constitute the third population of myoblasts and remain the sole source of DNA for postnatal muscle growth and repair (reviewed by Bischoff,1994).

The importance of satellite cells to postnatal muscle growth has been known for over two decades. Satellite cells are found in close association with myofibers between the plasmalemma and the basal lamina. They comprise about 30% of sublaminar nuclei at birth and decline to less than 5% in an adult (Bischoff, 1994). Microscopic examination of nuclei in two skeletal muscles from the pig showed an increase in the absolute number of satellite cells through the first 32 weeks of postnatal growth followed by a decrease in number from 32 to 64 weeks (Campion et al., 1981). In the same study, the number of myonuclei within the muscle fiber increased through the 64 weeks of age. Satellite cells can exit the cell cycle and remain quiescent until reactivated for hypertrophy or regeneration. When activated, satellite cells express either myf-5 or MyoD and undergo proliferation prior to fusion into existing or new muscle fibers, or they return to a quiescent state (Cooper et al., 1999). Therefore, postnatal muscle growth can involve hyperplasia of the mononucleated satellite cell which in turn supports hypertrophy in the multinucleated myofiber.

Satellite cells seemed to be a type of stem cell with the ability to proliferate and generate new satellite cells as well as differentiate and undergo fusion into myotubes. However, the discovery of a stem cell-like population in skeletal muscle that has hematopoetic potential (Jackson et al., 1999) has brought new insight into the specification of a satellite cell. *Pax7* is expressed by cultured satellite cells as well as by satellite cells within the muscle fibers. The role of Pax7 was subsequently determined by targeted gene disruption in mouse embryonic stem cells (Seale et al., 2000). Mice that lack a functional *Pax7* gene had essentially normal muscular development at birth but were significantly smaller than their control litter mates, failed to grow, and died within two weeks of age. Primary muscle cell preparations from Pax7 mutant animals produced only fibroblasts and adipocytes and no cells expressing myoblast markers such as desmin or myosin. However, the stem cell-like population was isolated from the Pax7 mutant mouse muscle. The muscle-derived stem cells from wild type mice have both myogenic and hematopoetic differentiation potential. However, muscle-derived stem cells from Pax7 mutant mice had enhanced hematopoetic potential and no myogenic potential. These results demonstrated that embryonic and fetal myoblasts which are induced to the myogenic lineage by Pax-3 have distinct progenitors that are different from satellite cells (Seale et al., 2000). Satellite cells require Pax-7 to induce expression of myf-5 or MyoD and specify the myogenic lineage. The presence of the stem cell-like population in the absence of Pax-7 could indicate that this population is the source of satellite cells in skeletal muscle (Seale et al., 2000). Understanding how this stem cell-like population of cells is activated and induced to differentiate into satellite cells will be important in understanding postnatal muscle growth and muscle fiber hypertrophy.

The total number of muscle fibers in an animal is generally determined by birth. Formation of primary myofibers does not appear to be a major source of variation for growth in domestic animals. Within a species, most animals have similar numbers of primary myofibers, regardless of birth weight, except in the extreme case of the runt pig, which has a small reduction in primary fiber number (Aberle,1984; Handel and Stickland,1987). The number of secondary myofibers are a source of variation in birth weight (Handel and Stickland,1987) and in the late stage of postnatal weight and average daily gain in pigs (Dryer et al.,1993). Growth potential between strains (Miller et al.,1975) and within a strain (Dryer et al.,1993) is also associated with the number of secondary myofibers. Furthermore, the nutritional plane of the dam can influence the number of secondary fibers during gestation (Dryer et al., 1994).

DISCOVERY OF MYOSTATIN

Numerous growth factors have been studied for their involvement in myogenesis and associated processes, including proliferation, differentiation, and fusion of primary myoblasts and immortalized muscle cell lines (rev. by Florini et al., 1996). Among the most studied extracellular signals are the insulin-like growth factors (IGFs), fibroblast growth factors (FGFs), and transforming growth factors (TGFs; reviewed by Florini et al., 1991). The ability of IGFs to stimulate both proliferation and differentiation of myoblasts is well established. On the other hand, FGFs stimulate proliferation but inhibit myogenin-induced terminal differentiation. Members of the TGF-ß family are negative growth factors for myogenesis that inhibit differentiation of transformed muscle cell lines as well as the proliferation and differentiation of primary satellite cells (Bischoff, 1994). The inhibition of proliferation is a common response of many cell types to TGF-ß in which activation of cyclin dependent kinase inhibitors p21 and p27 is blocked (reviewed by Massague et al., 2000). The inhibitory effects of TGF-ß on myoblast differentiation involves repression of the transcriptional activities of myogenin (Martin et al., 1992).

The identification of a novel TGF-ß member called GDF8/myostatin (McPherron et al 1997) has allowed characterization of a negative regulator of muscle growth. A full length cDNA, GDF-8, was isolated from a mouse skeletal muscle library and shown to be continually expressed throughout muscle development from mature somites to skeletal muscles of adults. The inhibitory role of GDF-8 on muscle development was clearly identified through targeted gene disruption using mouse embryonic stem cells. Loss of GDF-8 function resulted in an increase in muscle fiber numbers and muscle mass and the locus was renamed myostatin (McPherron et al., 1997).

DOUBLE MUSCLING IN CATTLE

Extreme muscling, referred to as double muscling, is reported in several breeds of cattle (Menissier, 1982). Animals expressing the double muscling phenotype are characterized by the presence of intermuscular grooves (Figure 1) and an increase in muscle mass of about 20% due to skeletal muscle hyperplasia. Carcasses produced from these animals are very desirable, with dramatic improvements in lean yield and tenderness (reviewed by Arthur 1995). However, the increase in muscle fiber number occurs *in utero*, producing a heavier fetus with abnormally large hips and shoulders and therefore, creating significant problems at birth (Menissier, 1982; Casas et al., 1999). The frequent need for Caesarian delivery has dampened enthusiasm for commercial exploitation of the phenotype.

Figure 1. Double muscling phenotype in a Belgian Blue animal.

The *mh* locus responsible for double muscling in Belgian Blue cattle has been mapped to the centromeric tip of bovine chromosome 2 (Charlier et al., 1995). Because of its phenotypic effect in mice is similar to double muscled cattle and because of its location on human chromosome 2q32, which is syntenic to bovine chromosome 2, myostatin was immediately recognized as a positional candidate gene for double muscling. Soon after the identification of myostatin, three independent groups (Grobet et al., 1997; Kambadur et al., 1997; McPherron and

Lee, 1997) characterized mutations in bovine myostatin that were associated with the double muscle trait. To date, six allelic sequence variants, all predicted to disrupt the function of the protein, explain the occurrence of double muscling in all but two double muscling breeds. The *nt821(del11)* mutation is an 11 bp deletion in the third exon, resulting in truncation of the bioactive carboxyterminal domain of the protein. This mutation has been found in Belgian Blues (Grobet et al., 1997; Kambadur et al., 1997; McPherron and Lee, 1997), Blonde d'Aquitaine, Limousin, Parthenaise, Asturiana and Rubea Gallega (Grobet et al., 1998). Another premature stop codon in the bioactive carboxyterminal domain of the protein is caused by a G to T transversion (*E291X*) and is found in the Marchigiana breed (Cappuccio et al., 1998). Three mutations producing premature stop codons are found in the N-terminal latency-associated peptide, including *nt419(del7-ins10)* which replaces 7 bp with an unrelated 10 bp and occurs in Maine-Anjou (Grobet et al., 1998), *Q204X* which is a C to T transition in the second exon, occurring in Charolais and Limousin (Grobet et al., 1998), and *E226X* which is a G to T transversion, also in the second exon, occurring in Maine-Anjou (Grobet et al., 1998). The *C313Y* mutation is a G to A transition, resulting in the substitution of a highly conserved cysteine with a tyrosine, disrupting the three-dimensional conformation of the bioactive carboxyterminal peptide. This mutation has been found in Piedmontese (McPherron and Lee, 1997) and Gasconne (Grobet et al., 1998). Grobet and colleagues also identified a conservative substitution (*F94L*), a silent substitution *(nt414(C-T))*, and four polymorphisms in intronic sequences, none of which were associated with the double muscling phenotype in the 10 breeds that were examined (Grobet et al., 1998). Clear associations between the loss-of-function variants and double muscling could not be demonstrated for Limousin and Blonde d'Aquitaine in that study, suggesting that either the *myostatin* gene is not involved or that additional, as-of-yet unidentified mutations in myostatin are responsible for double muscling in these breeds.

Allele-specific testing is now available for identifying myostatin mutations in cattle (Fahrenkrug et al., 1999; Karim et al., 2000), allowing characterization of the mode of inheritance. Prior models of inheritance included autosomal dominant, autosomal recessive, oligogenic and polygenic, reflecting the genetic heterogeneity of this trait in various breeds and the difficulty in classifying animals into discrete genotypes based on a continuous distribution for double muscling phenotype. Using genotyping systems, an autosomal recessive model has been established in Belgian Blues (Charlier et al., 1995) and Asturiana (Dunner et al., 1997). It has been previously proposed that double muscling arose from a single *mh* mutation in the Shorthorn breed that spread through Europe at the beginning of the 19th century (Menissier, 1982). However, the identification of a series of causative mutations in double muscled breeds suggests that a single origin is unlikely, although some degree of gene migration has occurred.

Genotyping systems have also allowed examination of phenotypic variation in animals of specific *mh* genotypes. While the extreme muscling characteristic of the double muscling phenotype is associated with the recessive *mh* allele and expressed in the homozygous *mh/mh* genotype, other traits such as calving difficulty and retail product yield, show an additive effect, with heterozygous +/*mh* animals being intermediate to the two homozygous conditions (Casas et al., 1998, 1999).

FURTHER CHARACTERIZATION OF MYOSTATIN

The specific role of myostatin in muscle development has not yet been determined. The phenotype of increased muscle fiber number in the myostatin deficient mouse and double muscle cattle indicate an effect on hyperplasia. However, the mechanism of muscle fiber number could be controlled through fetal myoblast proliferation or through terminal differentiation. The loss of myostatin activity could lead to prolonged mitogenic stimulation of myoblasts producing more founder myoblasts and fusion-competent myoblasts that form secondary muscle fibers. The loss of an inhibitor of terminal differentiation could result in the fusion of a higher proportion of the myoblasts than would normally participate in secondary fiber formation. As TGF-ß family members can potentially inhibit both proliferation and terminal differentiation, one or both processes may contribute to muscle fiber hyperplasia in myostatin deficient animals.

The discovery of myostatin has led to an initial series of experiment to its expression and potential function in normal muscle growth and maintenance of muscle mass. The highest levels of myostatin expression in cattle and pigs occurs during fetal development that coincides with muscle fiber formation followed by a decrease in expression near birth and early postnatal growth (Kambadur et al., 1997; Ji et al., 1998). The expression of myostatin was significantly elevated (65%) in the longissimus muscle of low birth weight pigs relative to normal littermates (Ji et al., 1998) indicating that myostatin expression may influence variation in birth weight. However, levels of myostatin RNA were not changed either by treatment with growth hormone or 3 days of food deprivation in growing pigs (Ji et al., 1998). Elevated levels of myostatin RNA were detected in double muscle calves relative to normal calves, indicating some type of feedback regulation of myostatin expression (Bass et al., 1999).

Two studies in the rat indicate that elevated myostatin expression was associated with a nonpathological loss of induced skeletal muscle mass. Immobilization of one of the hind limbs of rats to prevent them from bearing weight for 10 days resulted in 16% decrease in plantaris muscle mass accompanied by a 110% increase in myostatin mRNA and a 37% increase in myostatin protein (Wehling et al., 2000). Thirty minutes of daily weight loading on the limb were sufficient to prevent the loss of muscle mass but still resulted in increased myostatin mRNA. Similarly, rats exposed to microgravity during a 17 day space shuttle flight had 19-24% loss of muscle weight and myostatin/skeletal a-actin mRNA ratios that ranged from 2 to 5 fold higher than controls (Lalani et al., 2000). The level of myostatin protein was also elevated. After 13 days of acclimatization to earth's gravity, muscle mass and myostatin levels returned to normal. A study of the association of myostatin with weight-loss associated with chronic illness was conducted in men with the AIDS wasting syndrome (Gonzalez-Cadavid et al., 1998). Increased concentrations of immunoreactive myostatin was found in the muscles and serum of HIV-infected men with a reduced fat-free mass index relative to healthy men. These studies indicate that increased myostatin expression is associated with the loss of skeletal muscle. Therefore, myostatin has a role in the maintenance of muscle mass, in addition to early growth and development.

MUSCLE HYPERTROPHY IN SHEEP

The *callipyge* gene is a mutation in sheep responsible for pronounced muscle hypertrophy, primarily in muscles of the pelvic limb (Figure 2; Koohmaraie et al., 1995; Carpenter et al., 1996). Interestingly, the hypertrophy is absent at birth and develops only after approximately three weeks of age; thus, there is no increased risk of dystocia for callipyge lambs and no differences in birth weights between callipyge and normal lambs (Jackson et al., 1997a). While no differences are found in weaning weight or postweaning average daily gain, feed efficiency is improved and feed intake is reduced in callipyge lambs (Jackson et al., 1997a). There is some negative influences of the gene on wool traits, with a decrease of 12.7% in fleece weight and a decrease of 8.7% in staple length in callipyge ewes, when compared to wool production in normal ewes (Jackson et al., 1997a).

Figure 2. Conformation of a normal-appearing homozygous callipyge ram (left) and four callipyge offspring.

Muscles from lambs expressing the callipyge phenotype enlarge to differing degrees, and not all muscles are affected (Koohmaraie et al., 1995; Jackson et al., 1997c). The largest effect is found in the semimembranosus, with this muscle being 46% larger in callipyge lamb carcasses than in normal lambs. Enlargement of the muscles is primarily due to myofiber hypertrophy. Carpenter et al. (1996) found larger average diameters of the fast twitch, oxidative and glycolytic (FOG) and fast twitch, glycolytic (FG) muscle fibers for callipyge muscles, but smaller average diameter for the callipyge slow twitch, oxidative (SO) fibers. Fiber content was decreased 33.3% and 30.1% for callipyge SO and FOG fibers, respectively, while content of FG fibers increased 35.7%. Diameters of the three fiber types within the suprasinatus of callipyge and normal muscles did not differ. Increased muscle fiber diameter is evident in 5-week old (Lorenzen et al., 2000) but not in 2-week-old lambs (Carpenter and Cockett, 2000).

Callipyge animals are more desirable than normal animals for several carcass and meat characteristics (Koohmaraie et al., 1995; Jackson et al., 1997b). Dressing percentage and ribeye area are dramatically increased (about 30%) while all measures of fatness are decreased. Unfortunately, there is some concern with decreased tenderness of the loin (Kerth et al., 1995; Koohmaraie et al., 1995). This increased toughness seems to be limited to the callipyge loin and shoulder, with little or no effect on the leg. In consumer panel evaluations (Kerth et al., 1995), 27.9% more normal chops from the loin and 20.0% more normal chops from the shoulder were rated acceptable for tenderness than the corresponding callipyge chops. However, percentage of leg chops that were rated acceptable did not differ between callipyge and normal. Several postmortem tenderization methods are effective in improving the tenderness of callipyge lamb meat, including aging, electrostimulation, calcium chloride injection (Carpenter et al., 1997; Koohmaraie et al., 1998; Solomon et al., 1998), mechanical tenderization (Solomon et al., 1998) and freezing-thawing (Duckett et al., 1998).

The *callipyge* locus has been mapped to the distal end of ovine chromosome 18 (Cockett et al., 1994). Linkage between several microsatellite markers on OOV18 and *callipyge* was detected using 133 offspring produced from two callipyge males and 81 normal females. Additional characterization of the locus has demonstrated a unique mode of inheritance termed "polar overdominance" (Cockett et al., 1996; Freking et al., 1998), in which only heterozygous offspring inheriting the mutation from their sire express the phenotype. The three other genotypes are normal in appearance. Progeny data indicate that reactivation of the maternal *callipyge* allele occurs after passage through the male germ line, although this reactivation is not absolute (Cockett et al., 1996).

Experiments directed towards the positional cloning of *callipyge* are ongoing. Using bovine YAC and BAC libraries and an ovine BAC library, physical contigs of bovine (Shay et al., 2000) and ovine (Segers et al., 2000) origin that contain the *callipyge* gene have been constructed. The contigs span a 900 kb interval flanked by *IDVGA30* and *OY3* microsatellites. The locus has been more precisely mapped to a 450 kb region, using breakpoint mapping and additional markers isolated from these contigs (Berghmans et al., 2000). Two candidate genes, *DLK* and *GTL2,* have been localized to this region. Schmidt et al. (2000) demonstrated that these two genes are co-expressed in mice but respond in a reciprocal manner to loss of DNA methylation. *Dlk1*, a member of the epidermal-growth factor family that is paternally expressed, inhibits differentiation of preadipocytes into adipocytes. *Gtl2* encodes a non-translated RNA and is maternally expressed. Thus, *Dlk1* and *Gtl2* provide an example of coordinated imprinting of linked genes, similar to *Igf2* and *H19*. Preliminary work (Bidwell et al., 2000) showed altered postnatal expression of *GTL2* in muscles that undergo hypertrophy, suggesting the involvement of *GTL2* in the callipyge phenotype. However, the explanation of polar overdominance based just on *GTL2* expression was not obvious. Further studies are required to elucidate the involvement of *GTL2* and possibly *DLK* in callipyge.

Characterization of another gene responsible for an increase in the rib eye muscle of lambs has been recently reported (Banks, 1997). Effects of the *REM* locus are less dramatic than for *callipyge*, with an 11% increase in muscle mass limited to the *longissimus*. The *REM* locus appears to act as a dominant gene and has only minor effects on meat tenderness (McEwan et al., 2000). Using microsatellite markers from ovine chromosome 18, *REM* was localized to approximately the same position as *callipyge* (McEwan et al., 1998; Nicoll et al., 1998), suggesting that *REM* is allelic to *callipyge* or that mutations in closely linked genes are responsible for the two hypertrophy phenotypes.

Texel sheep are characterized by generalized muscular hypertrophy (Charlier and Leroy, 1996; Figure 3). Based on the association between mutations in the *myostatin* gene and double muscling in cattle, a study was recently initiated to investigate the possible involvement of *myostatin* in the hypertrophy of Belgian Texel sheep (Marcq et al., 1998, 2000). In this study, the entire coding sequence of *myostatin* was determined for the Texel and normally muscled Romanov controls; however, no sequence differences were identified. Interestingly, analysis of microsatellite markers flanking *myostatin* on chromosome 2 in a (Texel x Romanov) x Romanov backcross pedigree revealed significant linkage (Marcq, personal communication). Thus, the causative mutation responsible for hypertrophy found in Texel sheep may reside outside the coding segments of the *myostatin* gene.

Figure 3. Conformation of an eight-month old Belgian Texel lamb

CONCLUSION

The control of myogenesis involves carefully orchestrated expression of specific genes. Reprogramming of this process by naturally occurring genetic mutations are responsible for muscle enhanced traits in farm animals, such as double muscling and callipyge. Study of muscle hyperplasia and hypertrophy in livestock species will advance basic understanding of muscle formation.

REFERENCES

Aberle ED. Myofiber differentiation in skeletal muscles of newborn runt and normal weight pigs. *J Anim Sci* 1984; 59:1651-1656.

Arthur PF. Double muscling in cattle: a review. *Aust J Agric Res* 1995; 46:1493-1515.

Banks R The meat elite project: establishment and achievements of an elite meat sheep nucleus. *Proc Assoc Advance Anim Breed Genet* 1997; 12: 598-601.

Bennett AM, Tonks NK. Regulation of distinct stages of skeletal muscle differentiation by mitogen-activated protein kinases. *Science* 1997; 278:1288-1291.

Berghmans S, Segers K, Shay T, Georges M, Cockett N, Charlier C. Breakpoint mapping positions the *callipyge* gene within a 450 kilobase chromosome segment containing the *DLK* and *GTL2* genes. *Mamm Genome* 2000; (accepted).

Bidwell CA, Shay TL, Georges M, Beever JE, Berghmans S, Segers K, Charlier C, Cockett NE. Differential expression of the *GTL2* gene within the callipyge region of ovine chromosome 18. 2000; (in preparation).

Bischoff R. The satellite cell and muscle regeneration. In *Myogenesis*. Ed. Engel A G, Franszini-Armstrong C. New York, McGraw Hill, pp 97-118. 1994.

Buonanno A, Rosenthal N. Molecular control of muscle diversity and plasticity. *Devel Genet* 1996; 19:95-107.

Campion DR, Richardson RL, Reagan JO, Kraeling RR. Changes in the satellite cell population during postnatal growth of pig skeletal muscle. *J Anim Sci* 1981; 52:1014-1018.

Cappucio I, Marchitelli C, Serrachhioli A, Nardone A, Filippini F, Ajmone-Marsan P, Valentini A. A G-T transversion introduces a stop codon at the mh locus in hypertrophic Marchigiana beef subjects. *Proc., XXVIth Intern. Conf. Anim. Genet., Auckland, New Zealand* p. 78. 1998.

Carpenter CE, Cockett NE. Histology of longissimus muscle from 2-week-old and 8-week-old normal and callipyge lambs. *Can J Anim Sci* 2000; (in press).

Carpenter CE, Rice OD, Cockett NE, Snowder GD. Histology and composition of muscles from normal and callipyge lambs. *J Anim Sci* 1996; 74:388-393.

Carpenter, CE, Solomon MB, Snowder GD, Cockett NE, Busboom JR. Effects of electrical stimulation and conditioning, calcium chloride injection and aging on the acceptability of callipyge and normal lamb. *Sheep Goat Res J* 1997; 13:127-134.

Casas E, Keele JW, Fahrenkrug SC, Smith TPL, Cundiff LV, Stone RT. Quantitative analysis of birth, weaning, and yearling weights and calving difficulty in Piedmontese crossbreds segregating an inactive myostatin allele. *J Anim Sci* 1999; 77:1686-1692.

Casas E, Keele JW, Shackelford SD, Koohmaraie M, Sonstegard TS, Smith TPL, Kappes SM, Stone RT. Association of the muscle hypertrophy locus with carcass traits in beef cattle. *J Anim Sci* 1998; 76:468-473.

Charlier C, Coppieters W, Farnir F, Grobet L, Leroy PL, Michaux C, Mni M, Schwers A, Vanmanshoven P, Hanset R, Georges M. The *mh* gene causing double-muscling in cattle maps to bovine chromosome 2. *Mamm. Genome* 1995 6:788-792.

Charlier C, Leroy PL. Comparison of muscular fibers of double muscled Texel and Bleu du Maine. *Proc., 47th EAAP, Lillehammer, Norway,* Session S3.20. 1996.

Cockett NE, Jackson SP, Shay TL, Nielsen D, Moore SS, Steele MR, Barendse W, Green RD, Georges M. Chromosomal localization of the *callipyge* gene in sheep. *Proc Natl Acad Sci USA* 1994; 91:3019-3023.

Cockett NE, Jackson SP, Shay TL, Farnir F, Berghmans S, Snowder GD, Nielsen D, Georges M. Polar overdominance at the ovine *callipyge* locus. *Science* 1996; 273:236-238.

Cooper RN, Tajbakhsh S, Mouly V, Cossu G, Buckingham M, Butler-Browne GS. In vivo satellite cell activation via myf-5 and MyoD in regenerating mouse skeletal muscle. *J Cell Science* 1999; 112:2895-2901.

Cossu G, Borello U. Wnt signaling and the activation of myogenesis in mammals. *EMBO J* 1999; 18:6867-6872.

Duckett SK, Klein TA, Dodson MV, Snowder GD. Tenderness of normal and callipyge lamb aged fresh or after freezing. *Meat Sci* 1998; 49:19-26.

Dunner S, Charlier C, Farnir F, Brouwers B, Canon J, Georges M. Towards interbreed IBD fine mapping of the *mh* locus: double-muscling in the *Asturiana de los Valles* breed involves the same locus as the *Belgian Blue* cattle breed. *Mamm Genome* 1997; 8:430-435.

Dwyer CM, Fletcher JM, Stickland NC. Muscle cellularity and postnatal growth in the pig. *J Anim Sci* 1993; 71:3339-3343.

Dwyer CM, Stickland NC, Fletcher JM. The influence of maternal nutrition on muscle fiber number development in the porcine fetus and on subsequent postnatal growth. *J Anim Sci* 1994; 72:911-917.

Fahrenkrug SC, Casas E, Keele JW, Smith TPL. Technical note: Direct genotyping of the double-muscling locus (*mh*) in Piedmontese and Belgian Blue cattle by fluorescent PCR. *J Anim Sci* 1999; 77:2028-2030.

Freking BA, Keele JW, Beattie CW, Kappes SM, Smith TPL, Sonstegard TS, Nielsen MK, Leymaster KA. Evaluation of the ovine *callipyge* locus: I. Relative chromosomal position and gene action. *J Anim Sci* 1998; 76:2062-2071.

Florini JR, Ewton DZ, Magri KA. Hormones, growth factors, and myogenic differentiation. *Ann Rev Physiol* 1991; 53: 201-216.

Gonzalez-Cadavid NF, Taylor WE, Yarasheski K, Sinha-Hikim I, Ma K, Ezzat S, Shen R, Lalani R, Asa S, Mamita M, Arver S, Bhasin S. Organization of the human myostatin gene and expression in healthy men and HIV-infected men with muscle wasting. *Proc Natl Acad Sci USA* 1998; 95:14938-14943.

Grobet L, Martin L J R, Poncelet D, Pirottin D, Brouwers B, Riquet J, Schoeberlein A, Dunner S, Menissier F, Massabanda J, Fries R, Hanset R, Georges M. A deletion in the bovine myostatin gene causes the double muscled phenotypc in cattle. *Nature Genetics* 1997; 17:71-74.

Grobet L, Poncelet D, Martin LJR, Brouwers B, Pirottin D, Michaux C, Menissier F, Zanotti M, Dunner S, Georges M. Molecular definition of an allelic series of mutations disrupting the myostatin function and causing double-muscling in cattle. *Mamm Genome* 1998; 9:210-213.

Handel SE, Stickland NC. Muscle cellularity and birthweight. *J Anim Prod* 1987; 44:311-317.

Jackson KJ, Mi T, Goodell MA. Hematopoetic potential of stem cells isolated from murine skeletal muscle. *Proc Natl Acad Sci USA* 1999; 96:14482-14486.

Jackson SP, Green RD, Miller NF. Phenotypic characterization of Rambouillet sheep expressing the *callipyge* gene: I. Inheritance of the condition and production characteristics. *J Anim Sci* 1997a; 75:14-21.

Jackson SP, Miller MF, Green RD. Phenotypic characterization of Rambouillet sheep expressing the *callipyge* gene: II. Carcass characteristics and retail yield. *J Anim Sci* 1997b; 75:125-132.

Jackson SP, Miller MF, Green RD. Phenotypic characterization of Rambouillet sheep expressing the *callipyge* gene: III. Muscle weights and muscle weight distribution. *J Anim Sci* 1997c; 75:133-139.

Ji S, Losinski RL, Cornelius SG, Frank GR , Willis GM, Gerrard DE, Depreux FFS, Spurlock ME. Myostatin expression in porcine tissues: tissue specificity and developmental and postnatal expression. *Am J Physiol* 1998; 275: R1265-R1273.

Kambadur R, Sharma M, Smith TPL, Bass JJ. Mutations in myostatin (GDF8) in double muscled Belgian Blue cattle. *Genome Res* 1997; 7:910-916.

Karim L, Coppieters W, Grobet L, Valentini A, Georges M. Convenient genotyping of six myostatin mutations causing double-muscling in cattle using a multiplex olignucleotide ligation assay. *Anim Genet* 2000 (in press).

Kerth CR, Jackson SP, Miller MF, Ramsey CB. Physiological and sensory characteristics of callipyge sheep. *Texas Tech U. Res Rep., Tech Rep.* No. T-5-356, pp. 31-33. 1995.

Koohmaraie M, Shackelford SD, Wheeler TL, Lonergan SM, Doumit ME. A muscle hypertrophy condition in lamb (callipyge): characterization of effects on muscle growth and meat quality traits. *J Anim Sci* 1995; 73:3596-3607.

Koohmaraie M, Shackelford SD, Wheeler TL. Effect of prerigor freezing and postrigor calcium chloride injection on the tenderness of callipyge longissimus. *J Anim Sci* 1998; 76:1427-1432.

Lalani R, Bhasin S, Byhower F, Tarnuzzer R, Grant M, Shen R, Asa S, Ezzat S, Gonzalez-Cadavida NF. Myostatin and insulin-like growth factor –1 and –2 expression in the muscle of rats exposed to the microgravity environment of the neurolab spaceshuttle flight. *J Endocrinol* 2000; 167:417-428.

Lorenzen CL, Koohmaraie M, Shackelford SD, Jahoor F, Freetly HC, Wheeler TL, Savell JW, Fiorotto ML. Protein kinetics in callipyge lambs. *J Anim Sci* 2000; 78:78-87.

Marcq F, El Barkouki S, Elsen JM, Grobet L, Royo L, Leroy PL, Georges M. Investigating the role of myostatin in the determinism of double muscling characterizing Belgian Texel sheep. *Proc XXVIth Intern Conf Anim Genet Auckland New Zealand* p. 75. 1998.

Marcq F, Elsen J-M, Marot V, Bouix J, Coppieters W, Eychenne F, Laville E, Nezer C, Sayd T, Bibe B, Georges M, Leroy P. Mapping quantitative trait loci causing the muscular hypertrophy of Belgian Texel sheep. *Proc 50th EAAP Zurich Switzerland.* 1999.

Maroto M, Reshef R, Munsterberg AE, Koester S, Goulding M, Lassar AB. Ectopic Pax-3 activates MyoD and Myf-5 expression in embryonic mesoderm and neural tissue. *Cell* 1999; 89:139-148.

Martin JF, Li L, Olson EN. Repression of myogenin function by TGF-□1 is targeted at the basic helx-loop-helix motif and is independent of E2A products. *J Biol Chem* 1992; 267:10956-10960.

McEwan JC, Gerard EM, Jopson NB, Nicoll GB, Greer GJ, Dodds KG, Bain WE, Burkin HR, Lord EA, Broad TE. Localization of a QTL for rib-eye muscling on OOV18. *Proc XXVIth Intern Conf Anim Genet Auckland New Zealand.* p. 101. 1998.

McEwan JC, Broad TE, Jopson NB, Robertson TM, Glass BC, Burkin Hb, Gerard E M, Lord Ea, Greer GJ, Bain WE, Nicoll GB. Rib-eye muscling (REM) locus in sheep: Phenotypic effects and comparative genome localization. *Proc XXVIth Intern Conf Anim Genet Minneapolis MN USA*. p. 25. 2000.

McPherron AC, Lee SJ. Double muscling in cattle due to mutations in the myostatin gene. *Proc Natl Acad Sci USA* 1997; 94:12457-12461.

McPherron AC, Lawler AM, Lee SJ. Regulation of skeletal muscle mass in mice by a new TGF-ß superfamily member. *Nature* 1997; 387:83-90.

Menissier F. Present state of knowledge about the genetic determination of muscular hypertrophy or the double muscled trait in cattle. In: *Current Topics in Veterinary Medicine and Animal Science. Vol. 16: Muscle Hypertrophy of Genetic Origin and Its Use to Improve Beef Production.* Ed. King and Menissier, Martinus Nijhoff, p. 397-428. 1982.

Miller LT, Garwood VA, Judge MH. Factors affecting porcine muscle fiber type, diameter and number. *J Anim Sci* 1975; 41:66.

Nicoll GB, Burkin HR, Broad TE, Jopson NB, Greer GJ, Bain WE, Wright CS, Dodds KG, Fennessy PF, McEwan JC. Genetic linkage of microsatellite markers to the Carwell locus for rib-eye muscling in sheep. *Proc VI World Conf on Genetics Applied to Lvst Prod, Armidale, Australia*. 26:529. 1998.

Ruiz-Gomez M, Coutts N, Price A, Taylor MV, Bate M. Drosophila dumbfounded: a myoblast attractant essential for fusion. *Cell* 2000; 102:189-198.

Schmidt J, Matteson PG, Jones BK, Guan X-J, Tilghman SM. The *Dlk1* and *Gtl2* genes are linked and reciprocally imprinted. *Genes and Develop* 2000; 14:1997-2002.

Seale P, Sabourin LA, Girgis-Gabardo A, Mansouri A, Gruss P, Rudnicki MA. Pax7 is required for the specification of myogenic satellite cells. *Cell* 2000; 102:777-786.

Shay TL, Berghmans S, Segers K, Meyers S, Beever JE, Womack JE, Georges M, Charlier C, Cockett NE. Fine-mapping and construction of a bovine contig spanning the ovine *callipyge* locus. *Mamm Genome* 2000; (accepted).

Solomon MB, Carpenter CE, Snowder GD, Cockett NE. Tenderizing callipyge lamb with the hydrodyne process and electrical stimulation. *J Muscle Foods* 1998; 9:305-311.

Stockdale FE. Myogenic cell lineages. *Dev Biol* 1992; 154:284-298.

Segers K, Vaiman D, Berghmans S, Shay T, Beever J, Cockett N, Georges M, Charlier C. Construction and characterization of an ovine BAC contig spanning the *callipyge* locus. *Anim Genet* 2000; (in press).

Tajbakhsh S, Rocancourt D, Cossu G, Buckingham M. Redefining the genetic hierarchies controlling skeletal myogenesis: Pax-3 and Myf-5 act upstream of MyoD. *Cell* 1999; 89:127-138.

Walsh K, Perlman H. Cell cycle exit upon myogenic differentiation. *Curr Opin Genet Dev* 1997; 7:597-602.

Wehling M, Cai B, Tidball JG. Modulation of myostatin expression during modified muscle use. *FASEB J* 2000; 14:103-110.

Weintraub H. The MyoD family and myogenesis: Redundancy, networks and thresholds. *Cell* 1993; 75:1241-1244.

Chapter 10

THE HALOTHANE GENE, LEANNESS AND STRESS SUSCEPTIBILITY IN PIGS

Paramasivam Kathirvel and Alan L. Archibald
Roslin Institute (Edinburgh), Roslin Midlothian, EH25 9PS, Scotland, U.K.

Abstract Susceptibility to halothane-induced malignant hyperthermia is a feature of the porcine stress syndrome and is inherited as a monogenic recessive trait determined by the halothane locus (*HAL*). The stress susceptible HAL^n allele is associated with reduced body fat in pigs. There is compelling evidence that mutations in the gene encoding the skeletal muscle sarcoplasmic reticulum calcium release channel (also known as the ryanodine receptor, RYR1) are responsible for predisposition to malignant hyperthermia (MH) in pigs and humans. There is some limited evidence for associations between susceptibility to MH and body composition in humans. It has been proposed that the leakage of calcium from defective calcium release channels causes involuntary exercising and hence improved muscling and reduced fat in pigs. This explanation for the associations between HAL^n and reduced fat is currently more plausible than the alternative explanation of a fat/lean gene very closely linked to the *HAL/RYR1* locus.

INTRODUCTION

The *halothane* gene in pigs is associated with susceptibility to the porcine stress syndrome and with reduced levels of fat.

There are three phenotypes that are considered to be manifestations of a general porcine stress syndrome (PSS) - sudden (stress) deaths, pale soft exudative meat (PSE) and sensitivity to halothane-induced malignant hyperthermia (for reviews see MacLennan and Phillips, 1992; Archibald, 1991). The deaths arise from uncontrolled malignant hyperthermic reactions, which can be triggered by handling, sexual intercourse, excessive ambient temperature and a number of chemical agents. Man is also afflicted by halothane-induced malignant hyperthermia (MH). Although the human and porcine forms of this disorder are not identical,

J.B. Owen et al. (eds.), Animal Models – Disorders of Eating Behaviour and Body Composition, 173–190.

the similarities are sufficiently compelling to merit using porcine MH as a model for MH in humans.

Susceptibility to halothane-induced malignant hyperthermia in pigs has been shown to be controlled by a recessive gene at a single autosomal locus (*HAL*) with both alleles exhibiting variable penetrance. The identification of the mutation responsible for susceptibility to halothane-induced malignant hyperthermia by Fujii *et al.* (1991) is discussed later in this chapter.

Susceptibility to the porcine stress syndrome can be considered as a genetic disease in pigs. Tests for identifying pigs genetically predisposed to the porcine stress syndrome were developed. The initial test involved a controlled exposure to the anaesthetic gas halothane and was based on the assumption that susceptibility to halothane-induced malignant hyperthermia is predictive of susceptibility to other manifestations of the porcine stress syndrome. Subsequently, biochemical genetic markers known to be linked to the halothane (*HAL*) locus were also used for genetic diagnosis. As we explain later the halothane field test and the linked genetic markers have some limitations in the identification of pigs carrying the deleterious HAL^{n} allele.

Why, given the availability of tests for predicting susceptibility to this inherited disorder, did the incidence of the disease allele HAL^{n} increase in many breeds during the 1970s and 80s? The answer lies in the association between the halothane gene and leanness. Animals homozygous for the disease allele ($HAL^{n/n}$) were observed to be leaner than animals homozygous for the resistant allele ($HAL^{N/N}$). Thus, pig breeders who were selecting for reduced fatness were both wittingly and unwittingly increasing the frequency of the HAL^{n} allele. Pigs that are heterozygous for the disease allele ($HAL^{N/n}$) are generally resistant to porcine stress but have intermediate levels of fat when compared with the homozygous susceptible and resistant animals. Thus, such heterozygotes were often seen as desirable despite the fact that they are carriers of the disease allele.

In this chapter we describe the molecular genetic basis of the porcine stress syndrome in greater detail. In the context of this book on "Animal models of disorders of eating behaviour and body composition" we examine the question – is the "halothane" mutation responsible for both susceptibility to malignant hyperthermia and reduced levels of body fat.

Malignant hyperthermia and the halothane gene

Of the three manifestations of the porcine stress syndrome (sudden stress deaths, pale soft exudative meat and malignant hyperthermia) susceptibility to chemical induced malignant hyperthermia is the only one that can be studied in the living animal. Chemical-induction of malignant hyperthermia was first recognised in pigs over thirty years ago (Hall *et al.*, 1966; Harrison *et al.*, 1968). Subsequently, it became apparent that the breeds of pigs, that exhibited high frequencies of sudden (stress) deaths and of pale soft exudative (PSE) meat, were particularly prone to halothane-induced malignant hyperthermia (Harrison, 1972; Mitchell and Heffron, 1980).

Eikelenboom and Minkema (1974) developed a predictive field test for susceptibility to PSS based on a short controlled exposure to the anaesthetic gas halothane. Although exposure to halothane can induce malignant hyperthermia and death, if the anaesthetic mask is removed immediately the first signs of muscle rigidity are observed, susceptible individuals generally recover. The development of this non-lethal test was critical to subsequent studies of the genetic control of the disease.

Sensitivity to halothane-induced malignant hyperthermia has been shown to be controlled by a recessive allele (*n*) at a single autosomal locus (*HAL*) (Ollivier *et al.*, 1975; Smith and Bampton, 1977). Carden *et al.* (1983) tested two-locus models with a susceptibility locus and a suppressor locus in attempts to find a genetic explanation for the incomplete penetrance observed by Ollivier *et al.* (1975) and Smith and Bampton (1977). Carden *et al.* (1983) concluded that a single and strictly recessive mode of inheritance may not be enough to explain their observations in the ABRO British Landrace lines selected for high and low incidence of halothane sensitivity. After further studies on these lines, Southwood *et al.* (1988) concluded that the degree of penetrance had been underestimated and that a single locus model provided the most appropriate explanation of halothane sensitivity.

As halothane sensitivity is a recessive trait, the standard halothane challenge test fails to distinguish between the homozygous resistant animals ($HAL^{N/N}$) and heterozygous carriers ($HAL^{N/n}$). Progeny testing, which offers the only reliable way of genotyping animals, is both expensive and time consuming. Therefore, there was a need for a cheap and accurate test that would facilitate the unambiguous identification of all three genotypes - *N/N*, *N/n* and *n/n*.

By segregation analysis several biochemical genetic markers linked to the *HAL* locus were identified (for a review see Archibald and Imlah, 1985). In order to use linked genetic markers it is necessary to establish the phase of the linkage between alleles at the marker loci and alleles at the locus of interest. For the *HAL* locus this requires both halothane testing and marker genotyping parents and their offspring. To establish the linkage phase for the recessive HAL^n allele, a minimum of one halothane sensitive parent or offspring is needed. Once the haplotypes for any individual have been established, the haplotypes of its offspring can be deduced from their marker genotypes. A number of the deduced haplotypes will be in error, as a result of recombination disrupting the parental haplotypes. The genetic distance (recombination frequency) between the marker loci and the disease locus will determine this error rate. These genetic markers, therefore do not offer a direct test of the *HAL* genotypes of randomly chosen individuals. Such a direct test was only possible once the defect in the HAL^n gene product or the underlying mutation in the gene had been identified.

The molecular basis of susceptibility to halothane-induced malignant hyperthermia

The halothane challenge test fails to distinguish between homozygous normal pigs ($HAL^{N/N}$) and carriers ($HAL^{N/n}$). A DNA test for the causal mutation would solve this problem. Therefore, in the 1980s several groups started searches for the *HAL* gene (Archibald, 1987). The ryanodine receptor (*RYR1*) gene was identified as the *HAL* gene by a combination of positional candidate gene cloning, physiological candidate gene cloning, comparative gene mapping and good luck.

Physiological approaches

As muscle is the site of the phenotypic defects associated with malignant hyperthermia (MH) this genetic disorder is classified as a myopathy.

Ca^{2+} is an important regulator of muscle contraction, relaxation and energy metabolism. In halothane sensitive pigs increased calcium has been observed in the skeletal muscle cytoplasm. This abnormality in Ca^{2+} regulation is believed to be the primary cause of MH (Louis *et al.*, 1990). In skeletal muscle, there are essentially three membrane systems that regulate cytoplasmic calcium – the plasma membrane, mitochondira and the sarcoplasmic reticulum (SR). Louis *et al.* (1990) argue that as halothane and dantrolene have no effect on calcium regulation by the plasma membrane in normal or MH muscle, changes in this membrane are unlikely to be responsible for MH. Several groups have studied mitochondrial systems in MH and normal pigs and humans. Louis *et al.* (1990) concluded that although mitochondria may contribute to MH they are unlikely to be the site of the primary defect. The sarcoplasmic reticulum and associated transverse tubule components were suspected to be the site of primary defect in MH (Louis *et al.*, 1990).

Differences in calcium-induced Ca^{2+} release have been found between normal and MHS (malignant hyperthermia susceptible) muscle (Endo *et al.*, 1983). Calcium-induced calcium release and halothane-induced calcium release in pigs sensitive to halothane (MHS) (Ohnishi *et al.*, 1983) showed significantly higher sensitivity to calcium and maximum rates of Ca^{2+} release than normal muscles (Ohnishi *et al.*, 1983). Mickelson *et al.* (1988) found no differences in Ca^{2+} - ATPase activity, calsequestrin content and calcium accumulation in SR vesicles isolated from MHS and normal pigs. MHS SR releases more calcium than normal SR after 5 seconds with or without the addition of ATP, Mg^{2+}, or caffeine and produce elevated twitch forces in MHS muscles (Mickelson *et al.*, 1988). This difference could be due to an abnormal regulation of SR calcium release *in vivo* resulting in an increased sarcoplasmic Ca^{2+} level during a twitch and increased tension production. From studies of both isolated heavy SR membranes and intact muscle bundles Mickelson *et al.*

(1988) suggested that alterations in the SR ryanodine receptor protein may be responsible for the abnormalities in calcium regulation in MHS muscle.

The (skeletal muscle) ryanodine receptor (RYR1) is a key component of the calcium release channel in the sarcoplasmic reticulum. The term 'ryanodine receptor' is somewhat misleading as the binding of the plant alkaloid – ryanodine – to the protein is a laboratory rather than a physiological phenomenon.

The RYR molecule consists of two main regions: a C-terminal channel region and a large amino-terminal in the cytoplasmic region that constitutes the foot structure spanning the junctional gap between the sarcoplasmic reticulum and the transverse tubule.

In skeletal muscle, the signal for Ca^{2+} release is initiated at the neuromuscular junction. An action potential then spreads over the plasma membrane and transversely into the fibre via the T-tubules. The signal is then transferred across the triad junction, and Ca^{2+} is released from the terminal cisternae. The process is referred to as electro-mechanical coupling or depolarisation-induced Ca^{2+} release (Fleischer and Inui, 1989). In striated muscle, the ryanodine receptor (RYR1, the calcium release channel in the sarcoplasmic reticulum), dihydropyridine receptor (DHPR, the voltage-gated L-type calcium channel in the transverse tube) and SR Ca^{2+}-ATPase (SERCA, the calcium pump in the SR) proteins mediate excitation-contraction coupling (E-C coupling). Contraction is preceded by a massive release of stored Ca^{2+} through RYR channels, which open in response to activation of voltage gated Ca^{2+} channels through the established interactions of action, myosin and troponin (Catterall, 1991).

Studies by Louis and others suggest that the causal defect in MH is in Ca^{2+}-induced Ca^{2+} release channel (Mickelson *et al.*, 1988, 1989; Fill *et al.*, 1990; Louis *et al.*, 1990). Sarcoplasmic reticulum vesicles and single calcium release channels have been purified from pigs of different malignant hyperthermia genotypes and studied in reconstituted lipid bilayer membranes. Calcium release channels from malignant hyperthermia susceptible (MHS) pigs are slower to close after a contractile event, thus raising cytoplasmic calcium (Mickelson *et al.*, 1988; Fill *et al.*, 1990). The peptides generated by trypsin digestion of the Ca^{2+} release channel proteins isolated from pigs of different *HAL* genotypes have been compared (Knudson *et al.*, 1990). Differences in the tryptic peptides were observed between the two *HAL* homozygous classes, with the heterozygotes having an intermediate pattern. The different patterns appear to correspond to different rates of trypsin cleavage, rather than polymorphisms at the cleavage sites and may represent subtle conformational differences between putative allelic forms of the calcium release channel. These data support the earlier conclusion that the MH lesion does not correspond to an absence of function or a null mutation. Rather MH seems to be caused by a qualitative difference in the performance of a Ca^{2+} regulatory system, which under stress crosses a physiological Rubicon.

The ryanodine receptor / calcium release channel has a critical role in the regulation of calcium movements in skeletal muscle. Thus, the *RYR1* gene can be considered a physiological candidate gene for a number of myopathies including malignant hyperthermia.

Map based approaches

The *HAL* gene was assigned to a linkage group comprising biochemical markers glucose phosphate isomerase (*GPI*), the H and S blood groups (*EAH*, *EAS* or *S(A-O)*), the erythrocyte-6-phosphogluconate dehydrogenase (*PGD*) and alpha-1B-glycoprotein(*A1BG*)] in the order *S(A-O)-GPI, HAL-EAH-A1BG-PGD* (for a review see - Archibald and Imlah, 1985). The *HAL* locus was subsequently mapped to chromosome 6 through *in situ* hybridization assignments for the two linked markers (*GPI* and *PGD*) (Davies *et al*., 1988; Yerle *et al*., 1990). Arguably, *HAL* was the first trait gene mapped in pigs. DNA-based markers in the pig *GPI* gene were developed for prediction of pig *HAL* genotypes by linkage analysis (Davies *et al*., 1988). The pig *GPI* gene could have been used as a starting point for the positional cloning of the *HAL* gene.

In humans, *GPI* maps to chromosome 19q13.1. As pig MH was considered to be homologous to human MH, McCarthy *et al*. (1990) tested Irish MH pedigrees for linkage to chromosome 19 markers. MH was shown to be linked to markers from the *GPI* region of chromosome 19 with a peak lod score of 5.65 for theta = 0 with the cytochrome P450 gene (*CYP2A*) in one family (McCarthy *et al*., 1990). The multi point linkage analysis of the chromosome 19q12-13.2 region showed the distance between *D19S9* and *APOC2* to be 20 centimorgan (cM) and that they flank the *MH* locus (McCarthy *et al*., 1990). These results supported the assumption that MH in humans and pigs were genetic diseases with similar underlying causes. As MH in humans has multiple causes it was fortunate that the Irish pedigrees had the chromosome 19 associated form of MH.

Serendipitously the human skeletal muscle ryanodine receptor (*RYR1*) gene was mapped to chromosome 19q13.1 by MacKenzie *et al*. (1990) at about the same time that a human *MH* locus had been mapped to chromosome 19. However, the major inherited disease of muscle that was known to map to the long arm of human chromosome 19 was Myotonic Dystrophy (MD). Therefore, MacKenzie *et al*. (1990) tested a restriction fragment length polymorphism (RFLP) marker for the *RYR1* gene for linkage to the myotonic dystrophy locus. The *RYR1* locus was sub-localised on 19q13.1 and mapped near the MD susceptibility locus in the following order: 19qcen-*RYR1-D19S8-D19S16-D19S37-APOC2-CKMM-MD-pEWRB1*-19qter (MacKenzie *et al*., 1990). Several recombinants between *RYR1* and *MD* excluded *RYR1* as the gene responsible for MD. The authors speculated that as the *RYR1* region in humans shows homology to the region in pigs containing the *HAL* gene, *RYR1* could be a candidate for the malignant hyperthermia gene.

Linkage studies were performed in MH families with DNA markers in the *RYR1* gene. Susceptibility to MH was shown to co-segregate with the *RYR1* markers with a lod score of 4.20 at a linkage distance of zero cM (MacLennan *et al.*, 1990). Therefore, MacLennan *et al.* (1990) identified the ryanodine receptor as a strong candidate for malignant hyperthermia.

Subsequently MacLennan and his colleagues compared full-length *RYR1* cDNA sequences from Canadian MH-susceptible (MHS) Pietrain and MH negative (MHN) Yorkshire pigs and found 18 single nucleotide polymorphisms (Fujii *et al.*, 1991). One of the polymorphisms, a replacement of C with T at nucleotide 1843 alters the amino acid sequence from Arg to Cys at position 615 (Arginine in MHN and Cysteine in MHS).

In humans, a mutation equivalent to the pig Arg615Cys mutation has been found in some MH patients. The Arg 614 Cys change found in some human MH families results from a C1840T mutation (Galliard *et al.*, 1991). Other mutations in the human *RYR1* gene have been found in other MH pedigrees (*e.g.*: Barone *et al.*, 1999; see review McCarthy *et al.*, 2000).

Linkage studies with the *RYR1* point mutation and *GPI* and *PGD* markers in Canadian and British Landrace strains supported the hypothesis that C1843T is the causal mutation for porcine MH (Fujii *et al.*, 1991; Otsu *et al.*, 1991). This hypothesis has been confirmed in Hampshire, Duroc, Landrace, Yorkshire breeds (Houde *et al.*, 1993).

As discussed above, Fujii *et al.* (1991) found 18 nucleotide differences between the *RYR1* cDNA sequences for a Pietrain $HAL^{n/n}$ and a Yorkshire $HAL^{N/N}$ pigs. Harbitz *et al.* (1992) compared the cDNA sequence from a Norwegian Landrace $HAL^{N/N}$ pig to these previously published sequences. At five of the single nucleotide polymorphism (SNP) sites identified by Fujii *et al.* (1991) the Norwegian Landrace pig had the Yorkshire allele and at eleven of the SNP sites the Pietrain allele. For some regions of the cDNA Harbitz *et al.* (1992) sequenced two or more cDNA clones and identified seven single nucleotide polymorphisms, two of which corresponded to nucleotide differences observed between the Pietrain and Yorkshire sequences of Fujii *et al.* (1991). The Norwegian Landrace mutation at C1840T corresponds to C1843T in Canadian breeds. This observation supports the hypothesis that the C1843T mutation in the *RYR1* gene is a causal mutation in porcine MH (Harbitz *et al.*, 1992). However, one C to G sequence variation at nucleotide 6274 has been detected which potentially replaces Pro^{2092} (German Landrace) with Ala (Norwegian Landrace) (Leeb *et al.*, 1993).

Direct DNA tests for the halothane mutation have been used to eliminate this disease gene from most commercial lines. However, the HAL^{n} allele has been deliberately retained in a controlled manner in some lines. The DNA-based tests have also been used in studies of the effect of halothane genotypes on performance.

Halothane gene and leanness

In the United Kingdom average backfat thickness measured at the P2 position in pigs has been reduced over the last 20 years from in excess of 20 mm to current levels of approximately 11 mm (MLC, 1997). Improvements in carcass leanness have been achieved through a combination of genetic selection and improved nutrition.

As the halothane gene or specifically the HAL^n allele has often been considered to be associated with reduced levels of fat. The support for this association ranges from the anecdotal to the results of scientific studies. Stress susceptible pigs (i.e $HAL^{n/n}$) animals are perceived to be lean and well muscled. In part, this perception probably arose from the observation that the breeds, such as the Pietrain, in which PSS was prevalent, were also lean and well muscled. As the HAL^n allele was known to be associated with, and has subsequently be shown to be causal with respect to, porcine stress syndrome the increased frequency of the HAL^n allele was most readily explained if it was also associated with desirable traits (Southwood *et al.*, 1988). Thus, there has been a longstanding interest in associations between *HAL* genotypes and performance.

For example, Simpson and Webb (1989) compared the growth and carcass traits of pigs in a halothane positive (HP) line, which was essentially 100% $HAL^{n/n}$, a halothane negative (HN) line that included 25-35% carrier $HAL^{N/n}$ animals as well as $HAL^{N/N}$ pigs and crosses between the lines. When given food *ad libitum* the crosses showed increased appetite and growth rates which also resulted in increased backfat thickness. These differences were not significant. However, comparison of fully dissected carcasses revealed significant differences and in particular that the $HAL^{n/n}$ pigs were leanest and the crosses were intermediate between the HP and HN lines. The most significant differences observed in this study could be attributed to whether the pigs were fed *ad libitum* or a restricted diet.

Prior to the discovery of the molecular nature of the HAL^n allele studies of the effects of *HAL* genotypes on performance, including leanness were hampered by the inability to resolve all three *HAL* genotypes reliably. The DNA test for the *T/C* alleles at position 1843 of the *RYR1* gene allows all three genotypes (*RYR1*: *C/C*, *C/T* and *T/T* or *HAL*: *N/N*, *N/n* and *n/n* respectively) to be determined unambiguously (Fujii *et al.*, 1991) and eliminates the 5% diagnostic error associated with the halothane challenge test in pigs (Otsu *et al.*, 1991). Thus, in our review of studies of the effect of *HAL* genotypes on leanness published since 1991 we have only considered those in which the DNA test has been used.

Comparison of pigs from a line selected for low backfat depth and high growth rate with unselected controls derived from a common genetic origin showed that the *HAL*^{n} (*RYR1 T*) allele is associated with reduced carcass fatness (McPhee *et al.*, 1994). The pigs were fed *ad libitum* and both selected and control lines were segregating at the *HAL* locus. The *HAL*^{n} (*RYR1 T*) allele was also associated with reduced appetite and growth rates.

Rempel *et al.* (1995) examined the effects of *HAL* genotypes and sex on a range of meat quality and carcass traits including leanness in six different breed groups. They concluded that the *HAL*^{n} (*RYR1 T*) allele had beneficial effects on lean and fat (i.e. it was associated with reduced fatness and increased lean or muscle) but undesirable effects on meat quality. However, the effects of breed and sex were as significant as the *HAL* genotype effects. All the male pigs were castrates (barrows) and as expected they had less lean and more fat, but grew faster than the females. In contrast, entire male pigs would normally be expected to have more lean and less fat than females.

Effects of *HAL* genotypes on leanness have not been observed in all studies. García-Macías *et al.* (1996) found no significant effects of *HAL* genotypes on carcass fat and lean although they did find breed effects on these traits. For example, Duroc alleles were associated with increased backfat. Only the *HAL*$^{N/N}$ and *HAL*$^{N/n}$ genotypes were present. In a further comparison of *HAL*$^{N/N}$ and *HAL*$^{N/n}$ genotypes Leach *et al.* (1996) concluded that *HAL*$^{N/n}$ pigs had superior fat-free lean content.

Although lean content was increased in *HAL*$^{n/n}$ pigs and intramuscular fat content decreased as the number of *HAL*^{n} alleles increased the latter differences were not significant (de Smet *et al.*, 1996). Unfortunately, breed, halothane genotype and environment (farm) were confounded in this study. In a later study de Smet *et al.* (1998) found that increasing numbers of *HAL*^{n} alleles (i.e. from *HAL*$^{N/N}$ to *HAL*$^{N/n}$ to *HAL*$^{n/n}$) were associated with decreased backfat levels and increased lean content. Again in a comparison of sexes the barrows grew faster than females but produced less lean and more fat.

Some of the effects associated with the porcine stress syndrome can be ameliorated by careful treatment of the animals prior to slaughter and appropriate handling of the carcasses post-slaughter. In a study by Pommier *et al.* (1998) some of these environmental factors were controlled variables along with differences in sex and *HAL* genotypes. The experimental population consisted of the offspring of Yorkshire-Landrace sows (*HAL*$^{N/N}$) and Duroc boars (*HAL*$^{N/n}$). Feed to gain ratios were significantly better (i.e lower) for the *HAL*$^{N/n}$ pigs. The carcasses from the *HAL*$^{N/n}$ pigs yielded more lean and overall carcass yields were also higher. Although the pigs subjected to greater stress in the pre-slaughter period produced greater carcass yields the meat showed deleterious characteristics including lower pH values. Thus, Pommier *et al.* (1998) confirmed earlier findings about the desirable effects of the *HAL*^{n} allele on carcass yield and fat levels but also the undesirable effects on meat quality.

The effects of the *HAL*^{n} allele on lean and fat were confirmed in a study of 446 Pietrain x Large White F_2 pigs (Larzul *et al.*, 1997). Carcass yield and lean to fat ratios increased from $HAL^{N/N}$ to $HAL^{n/n}$ with the heterozygotes being intermediate. In contrast to earlier studies (Leach *et al.*, 1996; García-Macías *et al.*, 1996) Larzul and co-workers found no evidence for an effect of slaughter weight on the differences between the genotypes.

The phenotypes associated with different *HAL* genotypes have mainly been studied at the gross or whole animal level – backfat depths, total carcass lean and fat yields, growth rates and feed conversion ratios. There have been few attempts to examine the physiology of animals of different *HAL* genotypes. The hypothalamic-pituitary-adrenal axis has been studied most recently in $HAL^{N/N}$ and $HAL^{N/n}$ boars (Weaver *et al.*, 2000). Basal ACTH and cortisol levels were lower in the heterozygotes in contrast to an earlier study by the same group in which no differences were observed (Schaefer *et al.*, 1990). A remote sampling procedure was used in the more recent study and as noted by the authors stress-induced cortisol responses could have masked any differences between genotypes in the earlier study. However, again as observed by Weaver *et al.* (2000) the differences in basal levels may be specific to intact boars as castrates were used in the earlier study. Carcasses from the heterozygous boars also exhibited increased lean yields and reduced fat depths.

The observed associations between *HAL* (*RYR1*) genotypes and leanness could be caused by the *RYR1* disease mutation or by variation at another closely linked gene. For example, known linked genes include *APOE* (apolipoprotein E) and the transforming growth factor beta 1 gene (*TGFB1)* gene which have effects on myogenesis and adipogenesis and the hormone sensitive lipase (*LIPE*) gene which has an affect on body fat content as noted by Otsu *et al.* (1991). Studies on German Landrace and Pietrain breeds have shown that MH genotypes are associated with secondary changes in lipid composition. The MHS animals had lower quantities of lipid, but these quantities were also influenced by breed and sex. In general, meat from females has more fat than meat from males (Hartmann *et al.*, 1997). Dual-energy x-ray absorptiometry analysis in different growth periods showed that for pigs at 30, 60 and 90 kg, *N/N* pigs have more fat and less lean than the *n/n* pigs, while the *N/n* pigs were intermediate with respect to both fat and lean percentages (Mitchell and Scholz, 1997; Lahucky *et al.*, 1997).

Is the halothane gene causal for leanness?

There are two competing explanations for the observed associations between the HAL^{n} (*RYR1 1843T*) allele and leanness. First, there could be direct and causal relationship between the HAL^{n} (*RYR1 1843T*) mutation and reduced fatness. MacLennan and Phillips (1992) have suggested that the calcium leaking out of the defective calcium release channel could cause muscle contractures in sedentary animals. In effect they are arguing that the susceptible pigs are exercising involuntarily – perhaps the ideal exercise routine for time-conscious citizens of the developed world. The alternative explanation is that the *HAL/RYR1* locus is tightly linked to a gene that determines fat levels. Given the structure of pig breeding populations such putative linkage disequilibrium between a fatness gene and the *HAL/RYR1* gene is plausible.

Although these competing explanations or hypotheses are difficult to test there are two types of experiments that could inform the debate. First, the halothane mutation could be created *de novo* and the resulting phenotype examined – a reverse genetics approach. Second, the region around the *HAL/RYR1* locus could be scanned for genes influencing fat/leanness.

The reverse genetics approach

The evidence that the 'halothane' mutation (C1843T) in the pig *RYR1* gene causes susceptibility to malignant hyperthermia and reduced levels of body fat is derived from association studies and segregation analyses. In order to *prove* that the mutant is *causal* of the disease and/or the reduced levels of body fat it is necessary to carry out experiments to create the mutant *de novo* and monitor the consequences i.e. a reverse genetics approach. Until very recently such experiments were not possible in pigs. The capability to produce pigs by nuclear transfer (Polejaeva *et al.*, 2000) combined with gene targeting through homologous recombination (McCreath *et al.*, 2000) now makes it possible to consider such reverse genetics experiments in pigs. However, such experiments are very expensive and examining the relationship between the *RYR1 1843T* mutation and susceptibility to malignant hyperthermia and / or leanness is unlikely to be the first priority for this emerging opportunity.

As an alternative, creating a homologous point mutation in the mouse *Ryr1* model may allow the impact of this mutation to be studied experimentally. Although we and others have produced transgenic mice in which the *Ryr1* gene has been disrupted, we have failed to produce a mouse carrying the homologous C1843T mutation.

In order to study the functions of the skeletal muscle ryanodine receptor gene, the mouse *Ryr1* gene has been disrupted experimentally (Takeshima *et al.*, 1994; Moore *et al.*, 1998). The *Ryr1* gene has been knocked-out in exon 2 (Takeshima *et al.*, 1994) and exon 10 (Moore *et al.*, 1998) by inserting a neomycin marker gene into the gene.

Mice that are homozygous for the *Ryr1* exon 2 knock-out die perinatally and show gross abnormalities of skeletal muscle (Takeshima *et al.*, 1994). Mice with only one copy of the disrupted gene are normal. The mouse *Ryr1* gene has also been disrupted by inserting an *HPRT* minigene into exon 17 (Hitchin *et al.*, unpublished data). We have examined the phenotypes of transgenic mice with the *Ryr1* exon 17 knock-out and confirmed the findings of Takeshima *et al.* (1994) – mice homozygous for the exon 17 disruption also die perinatally and resemble the homozygous exon 2 knock-out mice. The exon 10 knock-out of *Ryr1* has been maintained in cell lines rather than as transgenic mice (Moore *et al.*, 1998).

Transgenic mice with a targeted mutation in either exon 2 or exon 10 of the *Ryr1* gene (Takeshima *et al.*, 1994; Moore *et al.*, 1998) have triad junctions lacking the feet structure (*dyspedic*) in skeletal muscle (Takekura *et al.*, 1995; Protasi *et al.*, 1998) which is essential for both muscular maturation and E-C coupling. This function of Ryr1 cannot be substituted by other subtypes of the receptors (Takeshima *et al.*, 1994). Dyspedic skeletal muscle cells, which do not express Ryr1, lack E-C coupling and also have about a 30 fold reduction in L-type Ca^{2+} current density (Fleig *et al.*, 1996; Nakai *et al.*, 1996).

We have tried several times to replace the HPRT cassette inserted into exon 17 of the murine *Ryr1* gene with sequences homologous to the porcine 1843T mutation but with no success. Unfortunately, the mice in which the *Ryr1* gene has been disrupted in either exon 2 or exon 17 are not informative for determining whether there is a causal relationship between *RYR1* genotype and leanness. As yet, we have been unable to produce mice carrying the homologous *RYR1 1843T* mutation that might allow this putative relationship to be studied.

What is the evidence for genes linked to HAL with an effect on leanness?

In parallel with the Human Genome project, with its objective of elucidating the genetic blueprint for *Homo sapiens,* projects to map the equally complex genomes of farm animals have been initiated. These livestock genome mapping projects have been justified in terms of understanding the number and effect of genes influencing economically important traits. The regions of the genome influencing quantitative traits are termed 'quantitative trait loci' or QTL. A genome scan for QTL is implemented by typing markers which are spaced at intervals through the genome and which are polymorphic (variable) in the population being studied. The phenomenon of genetic linkage means that each marker can be used to follow the inheritance of a section of linked chromosome. Body composition including fat / leanness is one of the traits that have been studied in such QTL-mapping experiments.

The first QTL mapped in pigs included regions on chromosome 4 with effects on growth rate, fatness and intestinal length (Andersson *et al.*, 1994). Briefly, a three-generation pedigree was established in which two European Wild Boars were crossed with eight Large White sows to generate F_1 pigs that were crossed *inter se* to yield 200 F_2 offspring. These pigs were genotyped for 105 polymorphic genetic markers and growth rates were recorded from birth to 70 kg. After slaughter, fat levels and intestinal length were noted. The largest effects revealed in the subsequent analyses were attributable to a region of chromosome 4. Fatness, whether measured as the percentage of fat in the abdominal cavity or backfat thickness, appeared to be under the control of QTLs that mapped to the proximal end of chromosome 4. Marklund *et al.* (1999) confirmed the presence of the fat QTL on chromosome 4 by segregation analysis in a subsequent backcross.

Several other groups have reported QTLs for body composition traits including fat / leanness in pigs (Geldermann *et al.*, 1996; Walling *et al.*, 1998; Rohrer and Keele, 1998; de Koning *et al.*, 1999; Óvilo *et al.*, 2000). Of the three studies that represented complete genome scans (Andersson *et al.*, 1994; Rohrer and Keele, 1998; de Koning *et al.*, 1999) only de Koning *et al.* (1999) found any evidence of fatness QTL on chromosome 6, the chromosome to which the *HAL* locus maps. The results reported by Geldermann *et al.* (1996) and Óvilo *et al.* (2000) were solely concerned with chromosome 6.

Geldermann et al. (1996) reported the presence on chromosome 6 of QTL influencing carcass composition and fattening parameters. However, as the *RYR1 T / HAL*^{n} allele is segregating in this population these data merely confirm the association between the *RYR1 T / HAL*^{n} allele and leanness. The results do not discriminate between the two possible explanations – a causal role for the *RYR1 T / HAL*^{n} allele or linkage disequilibrium between the *RYR1 T / HAL*^{n} allele and a fat trait gene.

No chromosome 6 fat QTL were reported for the first complete genome scan in pigs (Andersson *et al.*, 1994). Subsequently, it was revealed that the *RYR1 T / HAL*^{n} allele was unexpectedly present in one of the founder Wild Boars. The impact of the *RYR1 / HAL* locus was deliberately excluded from the initial analysis. The Uppsala group reported later the effect of the *RYR1 T / HAL*^{n} allele on performance in their Wild Boar x Swedish Yorkshire cross (Lundström *et al.*, 1995). In the F_2 generation, in which carcass traits were recorded, carriers of the *RYR1 T / HAL*^{n} allele were leaner than *HAL*$^{N/N}$ pigs – there were no *HAL*^{n} homozygotes in the population. Again the presence of the *RYR1 T / HAL*^{n} allele means that this study cannot be used to resolve whether or not the *RYR1 T / HAL*^{n} allele has a direct influence on fat levels.

Óvilo *et al.* (2000) found evidence for quantitative trait loci on chromosome 6 influencing intramuscular fat and backfat thickness. These authors demonstrated that the *RYR1 T / HAL*^{n} allele was absent from their experimental populations. Thus, Óvilo *et al.* (2000) have provided the first convincing evidence for genes on chromosome 6, other than the halothane gene, with effects on fat and leanness. The QTL described by Óvilo *et al.* (2000) lies about 30 centiMorgans (cM) distal to the *RYR1* locus, which unfortunately is not mapped in this study. The authors note that there are a number of positional candidate genes in this region distal to *RYR1* including the genes encoding the heart fatty acid binding protein (*FABP3*) leptin receptor (*LEPR*). The precision with which QTL can be located in genome scans is limited. Thus, the location of the QTL described by Óvilo *et al.* (2000) should only be considered approximate or provisional. Nevertheless the fat QTL reported by Óvilo *et al.* (2000) are not coincident with the *RYR1* locus. The carcass composition QTL described by Geldermann *et al.* (1996) is coincident with the *RYR1* locus.

MH and fat/lean in humans

The observation of associations between the *HAL^n* / *RYR1 1843T* allele and reduced fat in pigs is consistent across several studies and populations. Whether the association is causal or not remains unproven. The association between the susceptibility to halothane-induced malignant hyperthermia and the *RYR1 C1843T* mutation in pigs or the *RYR1 C1840T* mutation in humans is well established and supports the conclusion that these are homologous genetic diseases and lesions. Of perhaps greater interest in the context of this book is whether there is any evidence for a homologous reduction in body fat in humans genetically predisposed to malignant hyperthermia. It has been observed that some humans that are susceptible to MH are muscular and lean (Britt, 1976; Harriman *et al.*, 1978). Campbell *et al.* (1982) have reported that body fat percentage was lower in MHS males than in controls. They also found that thigh muscle diameter was greater in MHS females than in the controls, but there was no evidence for an increase in total muscle mass.

More recently energy expenditure has been examined in MHS and MHN individuals (Freymond *et al.*, 2000). The assignment of subjects to MHS, MHN or MHE (MH-equivocal) was on the basis of a muscle biopsy halothane-caffeine contracture test. The genotypes of the subjects for the known MH mutations were not determined. No significant differences in energy expenditure or insulin action were detected.

Conclusions

The association between the halothane gene, or more correctly between the *RYR1 T* / *HAL^n* allele, and leanness has been confirmed in many independent studies. The discovery of the molecular basis of the *RYR1 T* / *HAL^n* allele means that more recent studies have avoided the genotyping ambiguities present in earlier studies.

There are two competing explanations for the association between the *RYR1 T* / *HAL^n* allele and leanness – a causal relationship or a chance association arising from strong linkage disequilibrium with a "lean" allele at a linked fat QTL. Although fat QTL have been found on chromosome 6 in the absence of the *RYR1 T* / *HAL^n* allele neither of the competing hypotheses have been satisfactorily tested or proved.

We currently favour the explanation of a causal relationship between the *RYR1 T* / *HAL^n* allele and leanness. The mechanism proposed by David MacLennan to explain this causal link is that calcium leaking through faulty calcium release channels (RYR1) leads to involuntary exercising (MacLennan and Phillips, 1992). If this conclusion does not inform our understanding of obesity in humans it may at least confirm the benefits of exercise.

REFERENCES

Andersson L, Haley CS, Ellegren H, Knott SA, Johansson M, Andersson K. Andersson-Eklund L, Edfors-Lilja I, Fredholm M, Hansson I, Håkansson J, Lundström K. Genetic mapping of quantitative trait loci for growth and fatness in pigs. *Science* 1994; 263:1771-4.

Archibald AL. "A molecular genetic approach to the porcine stress syndrome." In: *Evaluation and control of meat quality in pigs*, (eds) Tarrant PV, Eikelenboom G, Monin G, Martinus Nijhoff Publishers, Dordrecht. pp 343-57, 1987.

Archibald A, Imlah P. The halothane sensitivity locus and its linkage relationships. Anim Bld Grps Biochem Genet 1985; 16: 253-63.

Archibald AL. "Halothane-induced malignant hyperthermia in pigs." In: *Breeding for disease resistance in farm animals*. edited by JB Owen & RFE Axford, pp 449-66. Wallingford : CAB International. 1991.

Barone V, Massa O, Intravaia E, Bracco A, Martino AD, Tegazzin V, Cozzolino S, Sorrentino V. Mutation screening of the RYR1 gene and identification of two novel mutation in Italian malignant hyperthermia families. *J Med Genet* 1999; 36:115-18.

Britt BA. Malignant hyperthermia. *Mod Med Canada* 1976; 31:511.

Campbell IT, Ellis FR, Halsall PJ, Hogge MS. Anthropometric studies of human subjects susceptible to malignant hyperpyrexia. *Acta Anaesth Scand* 1982; 26:363-7.

Carden AE, Hill WG, Webb AJ. The inheritance of halothane susceptibility in pigs. *Genet Sel Evol* 1983; 15: 65-82.

Catterall WA. Excitation-contraction coupling in vertebrate skeletal muscle: a tale of two calcium channels. *Cell* 1991; 64:871-4

Davies W, Harbitz I, Fries R, Stranzinger G, Hauge JG. Porcine malignant hyperthermia carrier detection and chromosomal assignment using a linked probe. *Anim Genet* 1988; 19:203-12.

de Koning DJ, Janss LLG, Rattink AP, van Oers PAM, de Vries BJ, Groenen MAM, van der Poel JJ, de Groot PN, Brascamp EW, van Arendonk JAM. Detection of quantitative trait loci for backfat thickness and intramuscular fat content in pigs (*Sus scrofa*). *Genetics* 1999; 152:1679-90.

de Smet SM, Pauwels H, De Bie S, Demeyer DI, Callewier J, Eeckhout W. Effect of halothane genotype, breed, feed withdrawal and lairage on pork quality of Belgian slaughter pigs. *J Anim Sci* 1996; 74:1854-63.

de Smet S, Bloemen H, van de Voorde G, Spincemaille G, Berckmans D. Meat and carcass quality in two pig lines of different stress-susceptibility genotype and their crosses. *Anim Sci* 1998; 66:441-47.

Eikelenboom G, Minkema D. Prediction of pale, soft, exudative muscle with a non lethal test for halothane induced porcine malignant hyperthermia syndrome. Netherlands *J Vet Sci* 1974; 99:421-6.

Endo M, Yagi S, Ishizuka T, Horiuti K, Koga Y, Amaha K. Changes in the Ca-induced Ca release mechanism in the sarcoplasmic reticulum of the muscle from a patient with malignant hyperthermia. *Biomed Res* 1983; 4:83-92.

Fill M, Coronado R, Mickelson JR, Vilven J, Ma J, Jacobson BA, Louis CF. Abnormal ryanodine receptor channels in malignant hyperthermia. *Biophys J* 1990; 50: 471-5.

Fleig A, Takeshima H, Penner R. Absence of Ca^{2+} current facilitation in skeletal muscle of transgenic mice lacking the type 1 ryanodine receptor. *J Physiol* 1996; 496:339-45.

Fleischer S, Inui M. Biochemistry and biophysics of excitation-contraction coupling. *Ann Rev BioPhys BioPhys Chem* 1989; 18:333-64.

Freymond D, Dériaz O, Frascarolo P, Reiz S, Jéquier E, Urwyler A. In vivo whole-body resting energy expenditure and insulin action in human malignant hyperthermia. *Anesthesiology* 2000; 93:39-47.

Fujii J, Otsu K, Zorzato F, De Leon S, Khanna VK, Weiler JE, O'Brien PJ, MacLennan DH. Identification of a mutation in porcine ryanodine receptor associated with malignant hyperthermia. *Science* 1991; 253:448-51.

Galliard EF, Otsu K, Fujii J, Khanna VK, De Leon S, Derdemezi J, Britt BA, Duff CL, Worton RG, MacLennan D.H. A substitution of cysteine for arginine 614 in the ryanodine receptor is potentially causative of human malignant hyperthermia. *Genomics* 1991; 11:751-5.

Garcia-Macias JA, Gispert M, Oliver MA, Diestre A, Alonso P, Munoz-Luna A, Siggens K, Cuthbert-Heavens D. The effects of cross, slaughter weight and halothane genotype on leanness and meat and fat quality in pig carcasses. *Anim Sci* 1996; 63:487-96.

Geldermann H, Müller E, Beeckamnn Knorr C, Yue G, Moser G. Mapping of quantitative-trait loci by means of marker genes in F_2 generations of Wild Boar, Pietrain and Meishan pigs. *J Anim Breed Genet* 1996; 113:381-7.

Hall LW, Woolf N, Bradley JWP, Jolly DW. Unusual reaction to suxamethonium chloride. *Brit Med J* 1966; 2:1305.

Harbitz I, Chowdhary B, Thomsen PD, Davies W, Kaufman U, Kran S, Gustavsson I, Christensen K, Hauge, JG. Assignment of the porcine calcium release channel gene, a candidate for the malignant hyperthermia locus to the 6p11->q21 segment of chromosome 6. *Genomics* 1992; 8:243-8.

Harriman DG, Ellis FR, Franks AJ, Sumner DW. Malignant hyperthermia myopathy – an investigation of 75 families. In: *Second international symposium on malignant hyperthermia*. Eds. Aldrete JA, Britt BA. Grune and Stratton, New York. pp67-87. 1978.

Harrison GG. Pale, soft, exudative pork, porcine stress syndrome and malignant hyperpyrexia - an identity? *J Sth African Vet Assoc* 1972; 43:57-63.

Harrison GG, Biebuyck JF, Terblanche J, Dent DM, Hickman R, Saunders J. Hyperpyrexia during anaesthesia. *Brit Med J* 1968; 3:594-5.

Hartmann S, Otten W, Kratzmair M, Seewald MJ, Iaizzo PA, Eichinger HM. Influences of breed, sex, and susceptibility to malignant hyperthermia on lipid composition of skeletal muscle and adipose tissue in swine. *Am J Vet Res* 1997; 58:738-43.

Houde A, Pommier SA, Roy R. Detection of the ryanodine receptor mutation associated with malignant hyperthermia in pure-bred swine populations. *J Anim Sci* 1993; 71:1414-8.

Knudson CM, Mickelson JR, Louis CF, Campbell KP. Distinct immunopeptide maps of the sarcoplasmic reticulum Ca^{2+} release channel in malignant hyperthermia. *J Biol Chem* 1990; 265:2421-4.

Lahucky R, Christian LL, Kovac L, Stalder KJ, Bauerova M. Meat quality assessed ante- and post-mortem by different ryanodine receptor gene status of pigs. *Meat Sci* 1997; 47:277-85.

Larzul C, Le Roy P, Gueblez R, Talmant A, Gogue J, Sellier P, Monin G. Effect of halothane genotype (*NN, Nn, nn*) on growth, carcass and meat quality traits of pigs slaughtered at 95 kg or 125 kg live weight. *J Anim Breed Genet* 1997; 114:309-20.

Leach LM, Ellis M, Sutton DS, McKeith FK, Wilson ER. The growth performance, carcass characteristics, and meat quality of halothane carrier and negative pigs. *J Anim Sci* 1996; 74:934-43.

Leeb T, Schmolzl S, Brem G, Brenig B. Genomic organisation of the porcine skeletal muscle ryanodine receptor (RYR1) gene coding region 4624 to 7929. *Genomics* 1993; 18:349-54.

Louis CF, Gallant EM, Remple E, Mickelson JR. Malignant hyperthermia and porcine stress syndrome: a tale of two species. *Pig News Inform* 1990; 11:341-4.

Lundstrom K, Karlsson A, Hakansson I, Johansson M, Andersson L, Andersoon K. Production, carcass and meat quality traits of F2-crosses between European wild pigs and domestic pigs including halothane gene carriers. *Anim Sci* 1995; 61:325-31.

MacKenzie AE, Korneluk RG, Zorzato F, Fujii J, Phillips M, Iles D, Wieringa B, Leblond S, Bailly J, Willard HF, Duff C, Worton RG, MacLennan DH. The human rynaodine receptor gene: its mapping to 19q13.1, placement in a chromosome 19 linkage group, and exclusion as the gene causing myotonic dystrophy. *Am J Hum Genet* 1990; 46:1082-9.

MacLennan DH, Phillips MS. Malignant hyperthermia. *Science* 1992; 256:789-94.

MacLennan DH, Duff C, Zorzato F, Fujii J, Phillips M, Korneluk RG, Frodis W, Britt BA, Worton RG. Ryanodine receptor gene is a candidate for prediposition to malignant hyperthermia. *Nature* 1990; 343: 559-61.

McCarthy TV, Healy JMS, Heffron JJA, Lehane M, Deufel T, Lehmann-Horn F, Farrall M, Johnson K. Localization of the malignant hyperthermia susceptibility locus to human chromosome 19q12-13.2. *Nature* 1990; 343:562-64.

McCarthy TV, Quane KA, Lynch PJ. Ryanodine receptor mutations in malignant hyperthermia and central core disease. *Hum Mut* 2000; 15:410-17.

McCreath KJ, Howcroft J, Campbell KHS, Colman A, Schnieke AE, Kind AJ. Production of gene-targeted sheep by nuclear transfer from cultured somatic cells. *Nature* 2000; 405:1066-9.

McPhee CP, Daniels LJ, Kramer HL, Macbeth GM, Noble JW. The effects of selection for lean growth and halothane allele on growth performance and mortality of pigs in a tropical environment. *Livest Prod Sci* 1994; 38:117-23.

Marklund L, Nyström P-E, Stern S, Andersson-Eklund L, Andersson L. Confirmed quantitative trait loci for fatness and growth on pig chromosome 4. *Heredity* 1999; 82:134-41.

Mickelson JR, Gallant EM, Litterer LA, Johnson KM, Rempel WE, Louis CF. Abnormal sarcoplasmic reticulum ryanodine receptor in malignant hyperthermia. *J Biol Chem* 1988; 263:9310-5.

Mickelson JR, Gallant EM, Rempel WE, Johnson KM, Litterer LA, Jacobson BA, Louis CF. Effects of the halothane-sensitivity gene on sarcoplasmic reticulum function. *Am J Physiol* 1989; 257:C787-94.

Mitchell G, Heffron JJA. The occurence of pale, soft, exudative musculature in Landrace pigs susceptible and resistant to the malignant hyperthermia syndrome. *Brit Vet J* 1980; 136: 500-6.

Mitchell AD, Scholz A. Dual-energy x-ray absorptiometry (DXA) analysis of growth and composition of pigs of different ryanodine receptor genotypes. *Archiv Fur Tierzucht-Archives of Animal Breeding* 1997; 40:47-56.

MLC. *Pig yearbook.* Meat and livestock commission, Milton Keynes, England. 1997.

Moore RA, Nguyen H, Galceran J, Pessah IN, Allen PD. A Transgenic myogenic cell line lacking ryanodine receptor protein for homologous expression studies: reconstitution of Ry1R protein and function. *J Cell Biol* 1998; 140:843-51.

Nakai J, Dirksen RT, Nguyen HT, Pessah IN, Beam KG, Allen PD. Enhanced dihydropyridine receptor channel activity in the presence of ryanodine receptor. *Nature* 1996; 380:72-5.

Ohnishi ST, Taylor S, Gronert GA. Calcium-induced Ca^{2+} release from sarcoplasmic reticulum of pigs susceptible to malignant hyperthermia. *FEBS letts* 1983; 161:103-107.

Ollivier L, Sellier P, Monin G. Déterminisme génétique du syndrome d'hyperthermie maligne chez le porc de Piétrain. *Annales de Génétique et de Sélection Animale* 1975; 7:159-66.

Otsu K, Khanna VK, Archibald AL, MacLennan DH. Co-segregation in F1 generation backcrosses of porcine malignant hyperthermia and a probable causal mutation in the skeletal muscle ryanodine receptor gene. *Genomics* 1991; 11:744-50.

Óvilo C, Pérez-Enciso M, Barragán C, Clop A, Rodríguez C, Oliver MA, Toro MA, Noguera JL. A QTL for intramuscular fat and backfat thickness is located on porcine chromosome 6. *Mamm Genome* 2000; 11:344-6.

Polejaeva IA, Chen SH, Vaught TD, Page RL, Mullins J, Ball S, Dai YF, Boone J, Walker S, Ayares DL, Colman A, Campbell KHS. Cloned pigs produced by nuclear transfer from adult somatic cells. *Nature* 2000; 407:86-90.

Pommier SA, Pomar C, Godbout D. Effect of the halothane genotype and stress on animal performance, carcass composition and meat quality of crossbred pigs. *Can J Anim Sci* 1998; 78:257-64.

Protasi F, Franzini-Armstrong C, Allen PD. Role of ryanodine receptors in the assembly of calcium release units in skeletal muscle. *J Cell Biol* 1998; 140:831-42.

Rempel WE, Lu MY, Mickelson JR, Louis CF. The effect of skeletal muscle ryanodine receptor genotype on pig performance and carcass quality traits. *Anim Sci* 1995; 60:249-57.

Rohrer GA, Keele JW. Identification of quantitative trait loci affecting carcass composition in swine: I Fat deposition traits. *J Anim Sci* 1998; 76:2247-54.

Schaefer AL, Doornenbal H, Sather AP, Tong AK, Jones SD, Murray AC. The use of blood serum components in the identification of stress-susceptible and carrier pigs. *Can J Anim Sci* 1990; 70:845-55

Simpson SP, Webb AJ. Growth and carcass performance of British Landrace pigs heterozygous at the halothane locus. *Anim Prod* 1989; 49:503-9.

Smith C, Bampton PR. Inheritance of reaction to halothane anaesthesia in pigs. *Genet Res* 1977; 29: 287-92.

Southwood OI, Simpson SP, Webb AJ. Reaction to halothane anaesthesia among heterozygotes at the halothane locus in British Landrace pigs. *Genet Sel Evol* 1988; 20:357-66.

Takekura H, Nishi M, Noda T, Takeshima H, Franzini-Armstrong C. Abnormal junctions between surface membrane and sarcoplasmic reticulum in skeletal muscle with a mutation targeted to the ryanodine receptor. *Proc Natl Acad Sci USA* 1995; 92:3381-5.

Takeshima H, Iino M, Takekura H, Nishi M, Kuno J, Minowa O, Takano H, Noda T. Excitation-contraction uncoupling and muscular degeneration in mice lacking functional skeletal muscle ryanodine-receptor gene. *Nature* 1994; 369:556-9.

Walling GA, Archibald AL, Cattermole JA, Downing AC, Finlayson HA, Nicholson D, Visscher PM, Walker CA, Haley CS. Mapping of quantitative trait loci on porcine chromosome 4. *Anim Genet* 1998; 29:415-24.

Weaver SA, Dixon WT, Schaefer AL. The effects of mutated skeletal ryanodine receptors on hypothalamic-pituitary-adrenal axis function in boars. *J Anim Sci* 2000; 78:1319-30.

Yerle M, Archibald AL, Dalens M, Gellin J. Localization of PGD and TGF 1 to pig chromosome 6q. *Anim Genet* 1990; 21:411-7.

PART 5

Anorexia models

Chapter 11

THE ANOREXIA MOUSE

Jeanette E. Johansen and Martin Schalling
Center for Molecular Medicine, Neurogenetics Unit, Karolinska Institutet.

ABSTRACT The anorexia (*anx*) mutation is an autosomal recessive and lethal mutation located on mouse chromosome 2. It causes severe malnutrition in preweanling mice as well as some neurological disturbances. Histological analyses show no abnormalities in the gastrointestinal system and blood parameters are normal. However, there are several interesting alterations of the distribution of many appetite regulating peptides in the hypothalamus of mutant mice. Serum leptin levels are also altered in *anx/anx* mice. The *anx* mouse is a useful model of malnutrition and particularly for exploring the role of hypothalamic regulation of food intake.

INTRODUCTION

Over the past five years there has been a tremendous increase in the understanding of the genetic regulation of food intake, using monogenic rodent models of obesity. Several genes have been cloned that, when mutated, cause obesity in the mouse and rat. These genes and their products have unravelled new biochemical pathways involved in obesity. Some of these genes have been shown to be important for the regulation of food intake also in humans (see chapter 7).

There has been less of a focus on genetic models of anorexia. One reason might be that there are much fewer genetic rodent models that shift the regulation of food intake, satiety or metabolic turnover towards anorexia than obesity. A possible explanation for this could be that even a relatively mild anorectic phenotype could lead to malnutrition and/or death by starvation early enough in life to affect the number and viability of the offspring. To our knowledge there is only one mouse model for anorexia – the anorexia (*anx*) mouse.

J.B. Owen et al. (eds.), Animal Models – Disorders of Eating Behaviour and Body Composition, 193–203.

THE ANOREXIA MOUSE

Phenotypic characteristics

Lois Maltais and co-workers first described the lethal anorexia (*anx*) mutation in 1984. Mutant mice (*anx/anx*) are characterised by poor appetite; stomach contents are reduced compared to normal littermates at about postnatal day 5 and continue so until death (Maltais *et al.*, 1984). Interestingly, the daily changes in stomach content in *anx/anx* mice are very similar to that observed in normal littermates from birth to 20 days of age (Maltais *et al.*, 1984). These data indicate that *anx/anx* mice fail to properly regulate the amount of food consumed rather than failing to eat for other reasons. Other characteristics are reduced body weight, emaciated appearance and abnormal behaviour including body tremors, headweaving, hyperactivity and uncoordinated gait.

Anorexic mice can be recognised at about 5-8 days of age by a thinning of the neck and tail. Later on they are easily recognised by their growth failure and emaciated appearance (Figure 1) as well as their abnormal behaviour.

Figure 1. A 21 days old anorexic (*anx/anx*) mouse (bottom) and a normal (+/+) littermate (top).

A significant difference in body weight between *anx/anx* and normal sib controls occurs at postnatal day 9 (Maltais *et al.*, 1984). The animals die at the age of three to five weeks depending on the genetic background (Johansen, unpublished). The different life spans on different backgrounds suggest that modifying genes contribute to the phenotype.

Histological analysis show no abnormalities in the gastrointestinal system and blood parameters such as total RBC, hematocrits, haemoglobin and mean cell volume are within normal range (Maltais *et al.*, 1984). No organ abnormalities have been found using routine stained sections. However, there are several interesting alterations of peptide distribution in the hypothalamus of *anx/anx* mice.

Hypothalamic peptide distribution in the *anx/anx* mouse

The hypothalamus has long been known as an important site for the regulation of food intake. Histochemical studies on the *anx/anx* mouse have demonstrated abnormalities in the hypothalamic arcuate nucleus (Broberger *et al.*, 1997a; 1998a; 1999). This nucleus houses several neurochemically defined cell populations (Everitt *et al.*, 1986), among them two populations that seem to play antagonistic roles in energy balance control. One population express neuropeptide Y (NPY) (Allen *et al.*, 1989; Chronwall *et al.*, 1985; deQuidt and Emson, 1986) and agouti gene related protein (AGRP) (Shutter *et al.*, 1997; Hahn *et al.*, 1998; Broberger *et al.*, 1998b), both stimulators of food intake. The other population expresses two inhibitors of food intake, pro-opiomelanocortin (POMC) (Bloom *et al.*, 1978; Bloch *et al.*, 1978; Watson *et al.*, 1978) and cocaine and amphetamine regulated transcript (CART) (Kristensen *et al.*, 1998; Elias *et al.*, 1998). Both these cell populations receive peripheral signalling from fat tissue via the peptide hormone leptin (Elmquist *et al.*, 1999).

The stimulatory effect of NPY on food intake is well characterised and has been shown to act primarily on carbohydrate consumption (Stanely and Leibowitz, 1985). NPY is expressed in the arcuate nucleus, in cells that innervate the rest of the hypothalamus. One particularly important target area for regulation of food intake is the paraventricular nucleus (PVN) (Bai *et al.,* 1985). When administered centrally into the PVN even low doses of NPY have a powerful stimulatory effect on food intake (Clark *et al.,* 1984; Stanely and Leibowitz, 1985) making NPY the most potent inducer of food intake known to date. AGRP is an endogenous antagonist of the anorexigenic melanocortin peptides (Ollman *et al.,* 1997; Fan *et al.*, 1997). It is expressed by 95% of the NPY neurones in the arcuate nucleus (Broberger *et al.*, 1998a).

In *anx/anx* mice, the NPY/AGRP neurones are characterised by increased NPY-like and AGRP-like immunoreactivity in the cell body and decreased staining in terminals (Broberger *et al.*, 1997a; 1998a). Interestingly mRNA levels of both NPY and AGRP in the arcuate nucleus of *anx/anx* mice are normal. This suggests that both NPY and AGRP are accumulated in the arcuate cell bodies of these neurones. One possible explanation for this observation is a deficiency in axonal transport mechanisms. An alternative explanation could be that the *anx* mutation prevents these neurones from developing neurite extensions.

POMC is the precursor protein for the melanocortin peptides adrenocorticotropic hormone (ACTH) and α-melanocyte-stimulating hormone (α-MSH). Fasting decreases POMC mRNA levels (Brady *et al.,* 1990) and melanocortin treatment induces anorexia (Fan *et al.,* 1997; Grill *et al.,* 1998). CART encodes putative peptides that have potent appetite-suppressing activity (Kristensen *et al.*, 1998; Lambert *et al.*, 1998). It even blocks NPY-induced feeding responses in rats, when injected intracerebroventricularly (Kristensen *et al.*, 1998; Lambert *et al.*, 1998). There is evidence that the CART peptide may act partly through the feeding-regulating circuitry operated by NPY, melanocortins and leptin (Elias *et al.*, 1998; Kristensen *et al.*, 1998; Lambert *et al.*, 1998). There is much anatomical and pharmacological evidence that the balance between NPY and POMC transmission may be correlated to the level of food intake. The projections of the arcuate POMC/CART neurones have been shown to be linked and parallel to the arcuate NPY system (Csiffáry *et al.,* 1990; Zhang *et al.,* 1994a; Fuxe *et al.,* 1997; Broberger *et al.,* 1997b; 1998a). Furthermore, POMC neurones express the NPY receptors Y1R and Y5R (Broberger *et al.*, 1997b; Fuxe *et al.,* 1997) and activation of these two receptors stimulates food intake (Stanley *et al.,* 1992; Gerald *et al.,* 1996), possibly through inhibition of POMC signalling.

anx/anx mice display decreased levels of POMC, Y1R, Y5R and CART mRNAs in the arcuate nucleus (Figure 2) (Broberger *et al.,* 1999; Johansen *et al.,* 2000), as well as a decreased number of cell bodies and terminals, as determined by staining with ACTH and α-MSH antiserum (Broberger *et al.*, 1999). They also show a dramatic reduction of POMC-positive dendritic extensions of the POMC cells (Broberger *et al.*, 1999). Taken together with the inability of the NPY neurones to release NPY and AGRP one speculative explanation is that the POMC neurones need some molecule(s) secreted from the NPY cells to survive. In the absence of this signal the cells deteriorate and die. Alternatively, fewer cells develop a POMC phenotype in the absence of NPY/AGRP signalling.

CART is also expressed in the dorsomedial hypothalamic nucleus and lateral hypothalamic area (DMH/LHA) as other parts of the brain and periphery (Douglass *et al.*, 1995; Koylu *et al.*, 1997). The intensity of CART signal in labelled cells in DMH/LHA of *anx/anx* mice does not differ from the healthy littermates, whereas the number of labelled cells is decreased (Figure 2) (Johansen *et al.*, 2000). A possible interpretation of this finding is that at least two distinct populations of CART expressing cells may exist in the DMH/LHA, differing in their response to the regulatory mechanisms affected in the *anx/anx* mouse. Further studies dissociating the functions of different hypothalamic CART expressing populations will be needed to address the functional relevance of these findings.

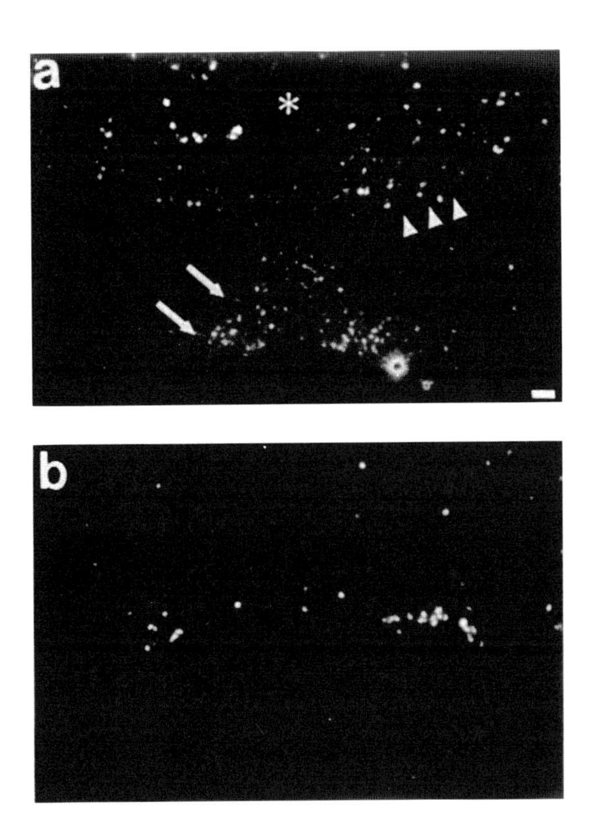

Figure 2. Darkfield micrographs of sections from the hypothalamus, arcuate nucleus (arrows) and the dorsomedial and lateral hypothalamic area (arrowheads) of control (a) and *anx/anx* (b) mice after hybridisation with CART probe. Note decreased CART mRNA expression in the arcuate nucleus and reduced number of CART expressing cells in the DMH/LHA of *anx/anx* mice. Asterisk indicates third ventricle. Scale bar indicates 100μm

The serotonergic system

The serotonergic system modulates several physiological functions including appetite and motor activity (Jacobs, 1991; Jacobs and Azmitia, 1992). Serotonin decreases total food consumption in a dose-dependent manner (Pollock and Rowland, 1981), and rat studies have shown that serotonin also has an effect on eating behaviour (Noach, 1994). Initial appetite seems to be unchanged, but satiety is obtained much quicker when rats are treated with serotonin (Noach, 1994). Furthermore, serotonin does not only decrease the total amount of food consumed but also alters the proportion of carbohydrates and protein therein. High levels of serotonin in the PVN result in relatively low carbohydrate and high protein intake and low levels have the opposite effect in rat (Shor-Posner *et al.*, 1986; Ashley *et al.*, 1979).

The abnormal behaviour of head weaving, body tremors, uncoordinated gait and hyperactivity of *anx/anx* mice is affected by serotonin (Maltais *et al.*, 1984). When treated with the serotonin precursor 5-hydroxy-DL-tryptophan 15 days old normal mice display the same type of abnormal behaviour as *anx/anx* mice. Similarly, 15 days old anorexic mice show body tremors and head weaving typical for anorexic mice 18 days of age or older, when treated with precursor 5-hydroxy-DL-tryptophan. Conversely, treating a 20 days old anorexic mouse with the serotonin antagonist 5, 7-dihydroxytryptamine, diminishes the severity of the neurological symptoms (Maltais *et al.*, 1984).

anx/anx mice have been shown to have an increase in the number and density of serotonergic fibres in the forebrain and the arcuate nucleus (Son *et al.*, 1994; Jahng *et al.*, 1998), which would be consistent with the experimental data on eating and motor behaviour.

Leptin and anorexia

Leptin is an adipocyte-derived hormone thought to regulate energy homeostasis through effects on food intake and thermogenesis (Campfield *et al.*, 1995; Halaas *et al.*, 1995; Pelleymounter *et al.*, 1995). Leptin was first discovered by Zhang and co-workers in 1994. Lack of leptin results in many phenotypic abnormalities including obesity, diabetes, infertility, stunted skeletal growth and impaired brain growth (Zhang *et al.*, 1994b). The main site of leptin action is the hypothalamus (Woods and Stock, 1996; Håkansson *et al.*, 1998; Elmquist *et al.*, 1998a; 1998b) where it regulates NPY, POMC and CART levels (Stephens *et al.*, 1995; Schwartz *et al.*, 1996; 1997; Kristensen *et al.*, 1998).

anx/anx mice have significantly reduced serum leptin levels from postnatal day 8 (Figure 3.) (Johansen *et al.,* In press). This may be expected as leptin is expressed in adipose tissue, of which *anx/anx* mice are almost completely devoid.

The low levels of leptin in postnatal *anx/anx* mice may add to some of the neurological abnormalities seen in these mice. Studies have shown a postnatal leptin surge at around the second postnatal week in mice independent of fat mass and food intake (Devaskar *et al.*, 1997; Ahima *et al.*, 1998). This peak in serum leptin levels is followed by the establishment of adult levels of corticosterone, estradiol and thyroxine (Ahima *et al.*, 1998) which may indicate that leptin expression is important for the development of the neuroendocrine axes independent of leptin's role in energy homeostasis. The mechanism for this possible additional role for leptin remains unknown, but it has been hypothesised that neonatal mice lack hypothalamic leptin receptors and is therefore insensitive to leptin inhibition of food intake (Devaskar *et al.*, 1997). However, further studies on the postnatal regulation of leptin are required in order to evaluate its putative developmental role in the *anx* mouse.

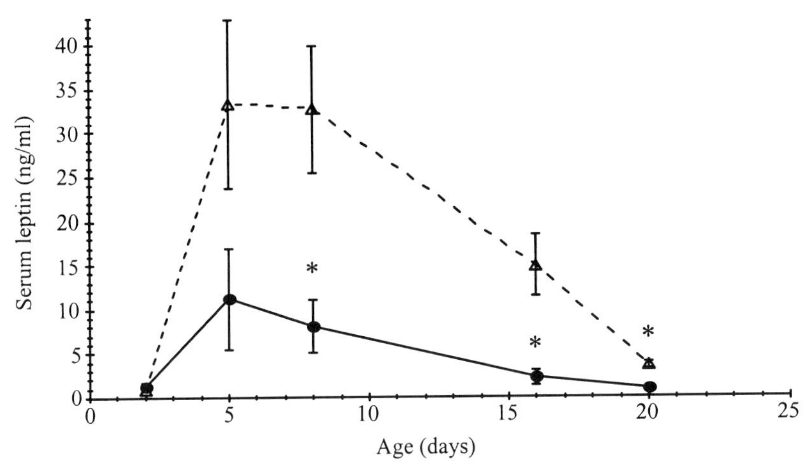

Figure 3. Serum leptin levels in *anx/anx* mice —●—, and normal littermates ---△---. Values represent means ± SEM; n = 5-8. * $P < 0.05$ comparing *anx/anx* with normal littermates using unpaired two-tailed t-test. Figure adapted from Johansen *et al.*, 2000.

Genetics of the *anx* mouse

The autosomal recessive *anx* mutation arose at the Jackson laboratory (Bar Harbor, ME, USA) in 1976 in the F_2 generation of a cross between DW/J and an inbred strain derived from a cross of *M. m. poschiavinus* to an inbred Swiss stock (Maltais *et al.*, 1984). The male carrier was crossed to a B6C3H-*a/a* F_1 female and the mutation has been maintained on this hybrid background at the Jackson laboratory. Linkage was found with the nonagouti locus, *a*, on chromosome 2. Two crosses and genotyping of approximately 5000 meioses have defined a 500kb interval for the *anx* gene. A contig has been generated across this region followed by a complete sequence of the interval. Genes within the interval are being analysed for alterations in the *anx* mouse.

Evidence for the existence of modifying genes come from a cross where the *anx* mutation was crossed with *M. m. castaneus* (CAST). *anx/anx* mice in the F_2 generation of the cross between B6C3H-*anx*/+ and CAST-+/+ do not start to show symptoms until around postnatal day 20 and they also survive longer, approximately five weeks as compared to three weeks on the B6C3H background (Johansen, unpublished).

Differential display analysis has identified a large number of genes for which the level of expression is markedly altered between *anx/anx* and wildtype mice. This information will contribute to elucidating the molecular pathways involved in the response to the deficit imposed by the *anx* mutation.

CONCLUSION

The *anx* mutation is linked to marked alterations in the hypothalamus, in particular in the histochemical distribution of signal substances known to have a potent regulatory role in control of food intake. The identification of the *anx* gene and its product(s) will add an important piece to the puzzle of understanding the regulation of food intake. The fact that the *anx* mutation affects many of the known hypothalamic appetite regulatory molecules makes the *anx* gene an interesting target for development of new tools and/or drug pharmaceuticals for the treatment of eating disorders. Mapping of modifier genes may further contribute to our understanding of the fine tuned control of appetite.

REFERENCES

Ahima RS, Prabakaran DJ, Flier S. Postnatal leptin surge and regulation of circadian rhythm of leptin by feeding. *J Clin Invest* 1998; 101: 1020-1027.

Allen YS, Adrian TE, Allen JM, Tatemoto K, Crow TJ, Bloom SR, Polak JM. Neuropeptide Y distribution in the rat brain. *Science* 1983; 221: 877-879.

Ashley DVM, Coscina DV, Harvey Anderson G. Selective decrease in protein intake following brain serotonin depletion. *Life Sci* 1979; 24: 973-984.

Bai FL, Yamano M, Shiotani Y, Emson PC, Smith AD, Powell JF, Tohyama M. An arcuato-paraventricular and –dorsomedial hypothalamic neuropeptide Y-containing system which lacks noradrenaline in the rat. *Brain Res* 1985; 331: 172-175.

Bloch B, Bugnon C, Fellmann D, Lenys D. Immunocytochemical evidence that the same neurones in the human infundibular nucleus are stained with anti-endorphins and antisera of other related peptides. *Neurosci Lett* 1978; 10: 147-152.

Bloom F, Battenberg E, Rossier J, Ling N, Guillemin R. Neurones containing beta-endorphin in rat brain exist separately from those containing enkephalin: Immunocytochemical studies. *Proc Natl Acad Sci USA* 1978; 75: 1591-1595.

Brady LS, Smith MA, Gold PW, Herkenham M. Altered expression of hypothalamic neuropeptide mRNAs in food-restricted and food-deprived rats. *Neuroendocrinology* 1990; 52: 441-447.

Broberger C, Johansen J, Schalling M, Hökfelt T. Hypothalamic neurohistochemistry of the murine anorexia (*anx/anx*) mutation: altered processing of neuropeptide Y in the arcuate nucleus *J Comp Neurol* 1997a; 387: 124-135.

Broberger C, Landry M, Wong H, Walsh J, Hökfelt T. Subtypes Y1 and Y2 of the neuropeptide Y receptor are respectively expressed in proopiomelanocortin and neuropeptide Y-containing neurones of the rat hypothalamic arcuate nucleus. *Neuroendocrinology* 1997b; 66: 393-408.

Broberger C, Johansen J, Johansson C, Schalling M, Hökfelt T. The neuropeptide Y/agouti gene-related protein (AGRP) brain circuitry in normal, anorectic, and monosodium glutamate-treated mice. *Proc Natl Acad Sci USA* 1998a; 95: 15043-15048.

Broberger C, De Lecea L, Sutcliffe J.G, Hökfelt T. Hypocretin/Orexin- and melanin-concentrating hormone-expressing cells form distinct populations in the rodent

lateral hypothalamus: Relationship to the neuropeptide Y and agouti gene-related protein systems *J Comp Neurol* 1998b; 402: 460-474.
Broberger C, Johansen J, Brismar H, Johansson C, Schalling M, Hökfelt T. Changes in neuropeptide Y receptors and pro-opiomelanocortin in the anorexia (*anx/anx*) mouse hypothalamus. *J Neurosci* 1999; 19: 7130-7139.
Campfield LA, Smith FJ, Guisez Y, Devos R, Burn P. Recombinant mouse OB protein: evidence for a peripheral signal linking adiposity and central neural networks. *Science* 1995; 269: 546-549.
Chronwall BM, DiMaggio DA, Massari VJ, Pickel VM, Ruggiero DA, O'Donohue TL. The anatomy of neuropeptide Y-containing neurones in rat brain *Neuroscience* 1985; 15: 1159-1181.
Clark JT, Kalra PS, Crowley WR, Kalra SP. Neuropeptide Y and human pancreatic polypeptide stimulate feeding behavior in rats. *Endocrinology* 1984; 115: 427-429.
Csiffáry A, Görcs TJ, Palkovits M. Neuropeptide Y innervation of ACTH-immunoreactive neurones in the arcuate nucleus of rats: a correlated light and electron microscopic double immunolabeling study. *Brain Res* 1990; 506: 215-222.
Devaskar SU, Ollesch C, Rajakumar RA, Rajakumar PA. Developmental changes in *ob* gene expression and circulating leptin peptide concentrations *Biochem Biophys Res Com* 1997; 238: 44-47.
deQuidt ME, Emson PC. Distribution of neuropeptide Y-like immunoreactivity in the rat central nervous system. Immunohistochemical analysis. *Neuroscience* 1986; 118: 545-618.
Douglass J, McKinzie AA, Couceyro P. PCR differential display identifies a rat brain mRNA that is transcriptionally regulated by cocaine and amphetamine. *J Neurosci* 1995; 15: 2471-2481.
Elias CF, Lee C, Kelly J, Aschkenasi C, Ahima RS, Couceyro PR, Kuhar MJ, Saper CB, Elmquist JK. Leptin activates hypothalamic CART neurones projecting to the spinal cord. *Neuron* 1998; 21: 1375-85.
Elmquist JK, Maratos-Flier E, Saper CB, Flier JS. Unraveling the CNS pathways underlying responses to leptin *Nat Neurosci* 1998a; 1: 445-450.
Elmquist JK, Bjorbaek C, Ahima RS, Flier JS, Saper CB. Distributions of leptin receptor mRNA isoforms in the rat brain *J Comp Neurol* 1998b; 395: 535-547.
Elmquist JK, Elias CF, Saper CB. From lesions to leptin: Hypothalamic control of food intake and body weight *Neuron* 1999; 22: 221-232.
Everitt BJ, Meister B, Hökfelt T, Melander T, Terenius L, Rökaeus Å, Theodorsson-Norheim E, Dockray G, Edwardson J, Cuello C, Elde R, Goldstein M, Hemmings H, Ouimet C, Walaas I, Greengard P, Vale W, Weber E, Wu J-Y, Chang K-J. The hypothalamic arcuate nucleus-median eminence Complex: Immunohistochemistry of transmitters, peptides and DARPP-32 with special reference to coexistence in dopamine neurones. *Brain Res Rev* 1986; 11: 97-155.
Fan W, Boston BA, Kesterson RA, Hruby VJ, Cone RD. Role of melanocortinergic neurones in feeding and the agouti obesity syndrome. *Nature* 1997; 385: 165-168.
Fuxe K, Tinner B, Caberlotto L, Bunnemann B, Agnati LF. NPY Y1 receptor like immunoreactivity exists in a subpopulation of β-endorphin immunoreactive nerve cells in the arcuate nucleus: a double immunolabelling analysis in the rat *Neurosci Lett* 1997; 225: 49-52.
Gerald C, Walker MW, Criscione L, Gustafson EL, Batzl-Hartmann C, Smith KE, Vaysse P, Durkin MM, Laz TM, Linemeyer DL, Schaffhauser AO, Whitebread S, Hofbauer KG, Taber RI, Branchek TA, Weinshank RL, A receptor subtype involved in neuropeptide-y-induced food intake *Nature* 1996; 382: 168-171.
Grill HJ, Ginsberg AB, Seeley RJ, Kaplan JM. Brainstem application of melanocortin receptor ligands produces long-lasting effects on feeding and bodyweight. *J Neurosci* 1998; 18: 10128-10135.
Hahn TM, Breininger JF, Baskin DG, Schwartz MW. Coexpression of Agrp and NPY in fasting-activated hypothalamic neurons. *Nat Neurosci* 1998; 1: 271-272.
Halaas JL, Gajiwala KS, Maffel M, Cohen SL, Chait BT, Rabinowitz D, Lallone RL, Burley SK, Friedman JF. Weight-reducing effects of the plasma protein encoded by the obese gene. *Science* 1995; 269: 543-546.

Håkansson ML, Brown H, Ghilardi N, Skoda RC, Meister B. Leptin receptor immunoreactivity in chemically defined target neurones of the hypothalamus. *J Neurosci* 1998; 18: 559-572.

Jacobs BL. Serotonin and behavior: emphasis on motor control *J Clin Psychiatry* 1991; 52 Suppl. 12: 17-23.

Jacobs BL, Azmitia EC. Structure and function of the brain serotonin system. *Physiol Rev* 1992; 72: 165-229.

Jahng JW, Houpt TA, Kim S-J, Joh TH, Son J.H. Neuropeptide Y mRNA and serotonin innervation in the arcuate nucleus of anorexia mutant mice. *Brain Res* 1998; 790: 67-73.

Johansen JE, Broberger C, Lavebratt C, Johansson C, Kuhar MJ, Hökfelt T, Schalling M. Hypothalamic CART and serum leptin levels are reduced in the anorectic (*anx/anx*) mouse. *Mol Brain Res* (In press).

Koylu EO, Couceyro PR, Lambert PD, Ling ND, DeSouza EB, Kuhar M.J. Immunohistochemical localization of novel CART peptides in rat hypothalamus, pituitary and adrenal gland. *J Neuroendocrinol* 1997; 11: 823-833.

Kristensen P, Judge MJ, Thim L, Ribel U, Christjansen KN, Wulff BS, Clausen JT, Jensen PB, Madsen OD, Vrang N, Larsen PJ, Hastrup S. Hypothalamic CART is a new anorectic peptide regulated by leptin. *Nature* 1998; 393: 72-76.

Lambert PD, Couceyro PR, McGirr KM, Dall Vechia SE, Smith Y, Kuhar M.J. CART peptides in the central control of feeding and interactions with neuropeptide Y. *Synapse* 1998; 29: 293-298.

Maltais LJ, Lane PW, Beamer WG. Anorexia, a recessive mutation causing starvation in preweanling mice. *J Hered* 1984; 75: 468-472.

Noach EL. Appetite regulation by serotoninergic mechanisms and effects of d-fenfluramine. *Netherlands J Med* 1994; 45: 123-133.

Ollmann MM, Wilson BD, Yang Y-K, Kerns JA, Chen Y, Gantz I, Barsh GS. Antagonism of central. melanocortin receptors in vitro and in vivo by agouti-related protein. *Science* 1997; 278: 135-138.

Pelleymounter MA, Cullen MJ, Baker MB, Hecht R, Winters D, Boone T, Collins F. Effects of the *obese* gene product on body weight regulation in *ob/ob* mice. *Science* 1995; 269: 540-543.

Pollock JD, Rowland N.E. Peripherally administered serotonin decreases food intake in rats. *Pharmacol Biochem Behav* 1981; 15: 179-183.

Schwartz MW, Baskin DG, Bukowski TR, Kuijper JL, Foster D, Lasser G, Prunkard DE, Porte Jr D, Woods SC, Seeley RJ, Weigle DS. Specificity of leptin action on elevated blood glucose levels and hypothalamic neuropeptide Y gene expression in *ob/ob* mice. *Diabetes* 1996; 45: 531-535.

Schwartz MW, Seeley RJ, Woods SC, Weigle DS, Campfield LA, Burn P, Baskin DG. Leptin increases hypothalamic pro-opiomelanocortin mRNA expression in the rostral arcuate nucleus. *Diabetes* 1997; 46: 2119-2123.

Shor-Posner G, Grinker JA, Marinescu C, Brown O, Leibowitz SF. Hypothalamic serotonin in the control of meal patterns and macronutrient selection. *Brain Res Bull* 1986; 17: 663-671.

Shutter JR, Graham M, Kinsey AC, Scully S, Lüthy R, Stark KL. Hypothalamic expression of ART, a novel gene related to agouti, is up-regulated in obese and diabetic mutant mice. *Genes Dev* 1997; 11: 593-602.

Son JH, Baker H, Park DH, Joh T.H. Drastic and selective hyperinnervation of central serotonergic neurones in a lethal neurodevelopmental mouse mutant, Anorexia (anx). *Mol Brain Res* 1994; 25: 129-134.

Stanley BG, Leibowitz SF. Neuropeptide Y injected in the paraventricular hypothalamus: A powerful stimulant of feeding behavior. *Proc Natl Acad Sci USA* 1985; 82: 3940-3943.

Stanley BG, Magdalin W, Seirafi A, Nguyen MM, Leibowitz SF. Evidence for neuropeptide Y mediation of eating produced by food deprivation and for a variant of the Y1 receptor mediating this peptide's effect. *Peptides* 1992; 13: 581-587.

Stephens TW, Basinski M, Bristow PK, Bue-Valleskey JM, Burgett SG, Craft L, Hale J, Hoffmann J, Hsiung HM, Kriauciunas A, MacKellar W, Rosteck Jr PR, Schoner B,

Smith D, Tinsley FC, Zhang X-Y, Heiman M. The role of neuropeptide Y in the antiobesity action of the *obese* gene product. *Nature* 1995; 377: 530-532.
Watson SJ, Akil H, Richard CW, Barchas JD. Evidence for two separate opiate peptide neuronal systems and the coexistence of beta-lipotropin, beta-endophin and ACTH immunoreactivities in the same hypothalamic neurones. *Nature* 1978; 275: 226-228.
Woods AJ, Stock MJ. Leptin activation in hypothalamus. *Nature* 1996; 381: 745.
Zhang X, Bao L, Xu Z-Q, Kopp J, Arvidsson U, Elde R, Hökfelt T. Localization of neuropeptide Y Y1 receptors in the rat nervous system with special reference to somatic receptors on small dorsal root ganglion neurones. *Proc Natl Acad Sci USA* 1994a; 91: 11738-11742.
Zhang Y, Proenca R, Maffei M, Barone M, Leopold L, Friedman J. Positional cloning of the mouse obese gene and its human homologue. *Nature* 1994b; 372: 425-432.

Chapter 12

ANOREXIA-LIKE WASTING SYNDROMES IN PIGS

S.C. Kyriakis
Clinic of Productive Animal Medicine, Faculty of Veterinary Medicine, Aristotle University of Thessaloniki, 54006, Thessaloniki, Greece

ABSTRACT In the last decades demand for high quality pork meat has led to the development of modern intensive pig production methods. As a result of the modification in the breeding, feeding, housing and management, modern intensively raised pigs have become more sensitive to social stress. The grouping imposed in many industrial pig units has consequences for feeding behaviour, feed intake, growth and the health status of pigs. Pigs, in particular those that have been bred for the purpose of extreme leanness, can develop irreversible self-starvation and emaciation. Anorexia in pigs develops mainly post-weaning as the wasting pig syndrome (WPS) or after farrowing as the thin sow syndrome (TSS). The clinical features of these syndromes show an uncanny resemblance to those of anorexia nervosa of humans. The aim of this chapter is to present WPS and TSS as possible animal models of disorders of eating and body composition in humans. WPS and TSS are related mainly to social and environmental stressors that occur during the very critical periods of lactation and weaning, and are widespread within some modern intensive pig husbandry systems

INTRODUCTION

During the last decades the demand for the production of high quality pork meat has risen worldwide, a result of the increase in the world population and greater consumer demands. This led to the development of modern intensive pig production methods, where the science and the practice of production has changed dramatically (Whittemore, 1998). Approximately one billion pigs are raised worldwide and pork is the dominant meat source, representing 40% of the total quantity of the meat consumed (Rothschild and Ruvinsky, 1997). As a result of the improvement and modification in the breeding, feeding, housing and

J.B. Owen et al. (eds.), Animal Models – Disorders of Eating Behaviour and Body Composition, 205–221.

management of pigs, modern intensively raised pigs are becoming more and more sensitive to social stress (Andersson, 1988).

The pig is a social animal by nature, but the grouping imposed in many industrial pig units are not those occurring in the wild life. It has been established that grouping has consequences on feeding behaviour, feed intake, growth and health status of the individual animal within the group (Morgan et. al., 1999). Pigs, in particular those that have been bred for the purpose of extreme leanness, can develop irreversible self starvation and emaciation (Treasure and Owen, 1997). Anorexia in pigs develops mainly post-weaning as the wasting pig syndrome or after farrowing as the thin sow syndrome (Kyriakis and Andersson, 1989, Kyriakis, 1991). The clinical features of these syndromes show an uncanny resemblance to those of anorexia nervosa (Treasure and Owen, 1997), a condition in which people, mostly women, would voluntary starve themselves (Gull, 1874).

In addition, the pig, because of its physiological and genetic similarities to man, serves as an excellent animal model for medical research (Rothschild and Ruvinsky, 1997). Pigs may be a useful model for studying the habits of eating in humans and / or digestive disorders because the species are roughly comparable in body weight and have a similar digestive process, being omnivorous and simple-stomached (Andersson, 1996). The aim of this chapter is to present the wasting pig syndrome (WPS) and the thin sow syndrome (TSS), both of which occur in modern intensively raised pigs, as possible animal models of disorders of eating and body composition in humans. WPS and TSS are related mainly to social and environmental stressors that occur during the very critical periods of lactation and weaning, and are widespread within some modern intensive pig husbandry systems (Treasure and Owen, 1997).

Before the actual presentation of WPS and, it is useful to consider some aspects of the genetics of behaviour, in particular feeding behaviour and the role of stress on growth and reproduction. These are essential to the understanding of underlying mechanisms involved in the appearance of both syndromes.

GENETICS OF BEHAVIOUR

The wild pigs of Asia and Europe were selected for domestication mainly because of their behaviour. While there was considerable variation in the behaviour of the pig within the species, the behavioural traits that favoured domestication included their omnivorous dietary needs, their relatively weak maternal-neonatal bonds, their precocial nature and their general adaptability (Ratner and Boice, 1976, Rothschild and Ruvinsky, 1997). It is known that the oldest breeds of domesticated European pigs and the present commercial pigs share many behavioural characteristics (Mitichashvili et. al. 1991). Over the years, looking at the species in terms of possible eating disorders, we can see a gradation from: (a) the primitive wild pig, which is omnivorous and with an apparently well controlled appetite mechanism, allied to a relatively lean body, to (b) the domesticated pig that was obese until the middle of the 20th century and to (c) thereafter, a progressively leaner animal bred to satisfy the modern consumer in developed economies (Owen, 1998).

The pig as a species, is known for its relatively copious feed intake. Pigs can reach feed intakes of 5% of their body weight per day, which exceeds the level of feed intake of most farm animal species (McGlone et. al., 1997, NRC 1998). If feed intake could be increased in young animals, it would improve the rate of weight gain and the feed efficiency. Thus, clearly there is a great economic reason to increase feed intake in the present domesticated lean lines of pigs. It is noticeable that the genetic basis for the control of the average daily feed intake (ADFI), has been less well studied than the closely correlated trait, average daily gain (ADG) (McGlone et. al., 1997). Furthermore, as a result of the intensive selection, which is applied among some modern lines of pigs for meat production, multiple selection goals may not result in increased feed intake, especially if ADG is emphasized. For example, with intensive selection for both less body fat and increased weight gain, feed conversion tends to improve, and interestingly, feed intake deteriorates (Vogeli, 1978, McGlone, 1997,).

The different problems of feed intake on commercial farms may imply genetic variation and thus, are amenable to change. One example is the adult sow whose feed intake is too high, which means wasting of feed and, if left unchecked, excessive body fat and body size. Limiting feed intake is the only solution to overcome this condition, but on the other hand it requires resources and alters the sow's behaviour (McGlone, 1997). This unusually high weight loss, especially when it takes place during lactation, is often followed by sow's failure to regain weight (Kyriakis, 1991). This condition is known worldwide as the thin sow syndrome (Kyriakis et. al, 1990, Kyriakis, 1991).

Another problem of feed intake occurs in the postweaning feeding period in young pigs. Pigs usually are not fond of dry feed in the hours shortly after weaning. Gradually, over the first few days and weeks after weaning, feed intake increases to reach the species' typical level (McGlone, 1997). Unfortunately, some pigs do not adapt in the new situation and develop a chronic stress syndrome. These pigs are known as wasting pigs or unthrifty pigs, and the pathological situation is known as the wasting pig syndrome (Kyriakis and Andersson, 1989).

Besides, changes in the feeding behaviour due to selection for modern lean pigs have possibly not only revived the 'primitive' leanness alleles, but also dredged up side effects known as the halothane gene (halothane-induced malignant hyperthermia). The halothane gene is associated with leanness, sensitivity to stress, and porcine stress syndrome. This condition can be observed in both pigs and humans. This gene, or the genes nearby, sensitizes animals to stress, sometimes affecting appetite and reproduction. The halothane mutation is due to a mutation in the calcium release channel (CRC) gene on chromosomal 6 of the pig, leading to a defect in calcium regulation (Fujji et. al., 1991). The symbol CRC is used for the locus encoding the skeletal muscle sarcoplasmatic reticulum, calcium release channel, also known as the ryanodine receptor (Andersson et. al 1994).

Body composition shows a high heritability based on the segregation of a single gene with major effects as happens with the halothane gene. On the other hand there is also a significant, though somewhat lower, level of heritability for the feed intake. Body composition is the primary controlled trait and feed intake variation is a dependent secondary effect (Owen, 1990). Finally, body composition is reviewed as a composite of several traits, each with their distinctive genetic basis, including major effects of genes at single loci. Studies involving twins, adopted offspring and other family relatives have demonstrated the high heritability (0.4-0.7) of body composition in many of the studies involved. Genotype-environment interactions with diet and activity occur in domesticated animals and humans and associations with voluntary choice of diet and level of activity are unfavourable. Body composition is the main reference for a normal homeostatic mechanism involving appetite and energy expenditure control (Owen, 1999).

FEEDING BEHAVIOUR

Growing pigs in modern production systems are kept in groups of similarly aged animals, sometimes of the same sex, in contrast to the "natural" groups, which included pigs of both sexes, young and old (Morgan et al, 1999). In particular, this stage of production takes place 3-4 weeks postpartum, when piglets are weaned (Whittemore, 1998). It has been recognized that such grouping can affect the production performance of individuals within the group (Morgan et. al, 1999). It is important to provide some information regarding the feeding behaviour of pigs housed within groups, because the wasting pig syndrome examined in this chapter is a condition observed in group-housed pigs. Moreover, the feeding behaviour forms the link between other social behaviours as well as between feed intake and growth.

The feeding behaviour of pigs housed within groups has been found to differ significantly from that of individually penned pigs. Within the social environment, growing pigs eat substantially fewer, but much larger meals than individually housed animals (De Haer and Merks, 1992). Besides, pigs within groups grow less quickly than we would expect, because our estimations are based on data from individually housed pigs (Morgan et. al., 1999). One possible explanation for the reduced growth rate is based on the proposition that pigs are more active in a group than when kept individually, because they interact with each other (Black, 1995) and the energy normally used for growth is therefore diverted to meet the demands of activity. Another explanation is that housing pigs in close proximity to each other within groups affects the thermal environment of the individual which may prevent effective loss of body heat. This results inreduction in feed intake. It is obvious that there are differences in the behavioural constraints related to the social environment, and in some ways social living affects the feed intake and growth.

THE ROLE OF STRESS ON GROWTH AND REPRODUCTION

Social stress is common on commercial pig units, mainly at the time of weaning of pigs and sows and during the first period of growth. Therefore, several stress induced diseases and stress related syndromes are now recognized in the pig. Before presenting the wasting pig syndrome and the thin sow syndrome, which are stress related, it is important to examine the role of stress on growth and reproduction (Kyriakis, 1989). In the terms of behaviour and welfare, stress occurs as the consequence of the restraint which is applied at intensive pig production units, and in particular, as it is seen in stall-housed pregnant and lactating sows and in common stocking densities among weaning and growing pigs (Whittemore, 1998).

Stress is an adaptive phenomenon, which represents a psycho-physiological reaction to the external stimuli or stressors (Kyriakis, 1989). The major stressors are disease, injury and subjection to aggression. Stress may also occur as a result of other environmental problems, for example at weaning, when piglets are removed from the sow to another building and they are mixed with other piglets (Whittemore, 1998). A stimulus becomes a stressor by affecting directly the individual through a sensory or metabolic process, which is stressful itself. Such stressors elicit a response via lower brain sensory mechanisms without involving higher interpretive brain centers. A stimulus can also become a stressor by virtue of the cognitive interpretation of meaning that the individual assigns to the stimulus. Thus, if an individual interprets or perceives some stimulus as being aversive or challenging, stress response will be elicited (Everby and Rosenfeld, 1981). In either case, the stress response is one of pituitary adrenal activation resulting in release of the pituitary and medullary tropic hormones ACTH, cortisol, epinephrine, norepinephrine and dopamine. Short bursts of adrenal cortical and medullary output are beneficially adaptive, but chronic hyperactivity of the pituitary-adrenal axis is regarded as maladaptive response. Stress has been linked to adrenocortical function and resultant acceleration of vascular aging in several species including the pig (Brambvell, 1965, Henny and Stephens, 1987).

In particular, one possible underlying mechanism that is responsible for the observed effects is the involvement of stress in growth regulation (Chapple, 1993). Reduced protein deposition, which results in growth retardation is due to the stress of the pig resulting from the need to maintain its social position within the group. This stress is acting via biochemical factors directly on tissue growth through growth hormone, cortisol insulin-like growth factor and cytokines (Morgan et. al., 1999). A reduction of circulating growth hormone could be expected to reduce protein deposition (MacRae and Lobley, 1991, Morgan et. al., 1999) although the mechanism by which stress should cause such a reduction of the growth hormone secretion is not clear. Cortisol increases proteolysis to supply amino acids for gluconeogenesis (Oliverio, 1987). Cytokines have also this effect and they additionally reduce the feed intake (Grimble, 1993). Both cortisol and cytokines also stimulate lipolysis. and chronic treatment of growing pigs with pharmacological doses of corticosteroid resulted in suppression of the growth rate (Oliverio, 1987, Grimble, 1993).

Another possible underlying route, through which socially induced stress might influence nutrient demand and growth rate, is through its influence on the animal's immune status. McGlone et. al. (1993), showed that low ranking pigs suffered a decrease in natural killer cell cytotoxicity following transport, in comparison to dominant pigs. Therefore, stressed pigs of low social status might be more susceptible to infections which in turn, could lead to decreased growth via clinical or subclinical pathogenic routes (Johnson and von Borell, 1994). In addition, pigs with a lesser activity of the immune system, due to medicated early weaning, have a lower ratio of T helper/T suppressor cytotoxic cells, a greater capacity for protein growth and, consequently higher amino acid requirements, than conventionally reared pigs. Furthermore, the administration of an inactivated Parapoxvirus immunomodulator (Baypamun), prevents wasting pig syndrome, which is stress induced (Kyriakis et. al., 1998). Baypamun triggers some non-antigen mediated mechanisms of the immune system such as interferon production, and cytotoxic activity of natural killer cells. Additionally it improves the reproductive performance of gilts (young breeding females) , mainly when they are exposed to transportation stress (Kyriakis et. al., 1996).

Finally, one outcome of high stress is the reduction of reproductive efficiency. Some manifestations of stress are reduced litter size, failure of oestrus, poor conception rates and sporadic attainment of puberty. In stressed young growing animals, the endocrine changes outlined augment their ability to cope with stress, but in a reproducing female there is a cascade of other hormonal events which follow. These secondary events can lead to ovarian dysfunction, complete infertility and embryonic death. Ovarian function depends upon a sequence of finely orchestrated endocrine phenomena and their disruption results in endocrine chaos. Adrenal hyperactivity is associated with asynchrony of the principal reproductive events. One mechanism by which stress influences reproduction is via ACTH, which directly suppresses pituitary response to gonadotrophin releasing hormone (GnRH) or decreases its secretion from the hypothalamus and this action is mediated by the adrenal gland (Varley and Stedman, 1994)

WASTING PIG SYNDROME (WPS)

Social and environmental stressors during the critical period of weaning frequently not only affect the performance but also the health status of the piglets. Pigs unable to cope with the changes associated with weaning, develop a condition characterized by retardation of growth and a listless appearance, known as the wasting pig syndrome (WPS) (Kyriakis and Andersson 1989). There are many possible negative factors (stressors) to which a pig can be exposed in connection with weaning, and they could be divided into different categories: a) the fact that weaning demands a nutritional change. Prior to weaning, piglets have access to hourly meals of high quality, readily digestive ingredients consisting of simple carbohydrates, proteins and fats provided by the sow's milk (Pond and Maner, 1984). Although piglets voluntarily consume solid feed in appreciable quantities they still rely on milk as an important food source for a long time (Algers et al., 1988). On the other hand piglets are forced to adapt to a different diet which leads to dramatic changes in their enzyme activity (Hefner, 1987). b) The need to utilize the feeders more may increase the frequency of agonistic interactions between littermates, particularly if feeding space is limited. and, c) There is also an effect because of the change of the environment, if the pigs are moved after weaning and d) the removal of the mother itself will probably cause a broad spectrum of various psychological strains (Algers et. al., 1988).

One of the most important negative factors that a pig is exposed to is the mixing of pigs after weaning. In modern pig practice, disturbances in the social hierarchy are purposely introduced. When pigs from different litters are put together for the first time they start fighting, and over the next 24 to 48 hours a new dominance hierarchy is established (Andersson, 1988). The consequences of fighting among pigs are severe. Aggression among newly weaned pigs mixed together into broken family groups can reduce the growth of all the pigs in the pen for a period of up to 2 weeks while the new hierarchy is established. If pigs of equal strength and dominance meet repeatedly, individual animals can be severely affected. Pigs demoted in the social order by fighting may subsequently fare less well, may be denied full access to feeders, and fail to thrive. These pigs are not only inefficient in their growth, but they are also susceptible to diseases and are therefore, a threat to the whole group (Whittemore, 1998).

During the period of rank order establishment, there is a strong activation of the pituitary-adrenal axis. As already mentioned short bursts of adrenal cortical and medullar output are beneficially adaptive but chronic hyperactivity of the pituitary-adrenal axis is regarded as a maladaptive response, which leads to elevated levels of circulating corticosteroids, depression of the immune system and secretion of catecholamine (Andersson et. al., 1986, Fabiansson, 1986). Because pens in modern pig housing do not offer hiding places or refuges during aggression, the pigs may stay in a state of constant stress during the period of rank order establishment. Clinical signs of a stress response following immediately after mixing of pigs are a reduction in the growth rate, feed intake and feed conversion efficiency. Within a few weeks, most of the pigs adapt to the new situation and social order, so that their feed utilization and body-weight increases are normalized. However, some pigs do not recover from the initial growth restraint and the inability of these pigs to cope with the new situation causes severe stress responses with serious health consequences. Those animals become wasting piglets, a chronic form of the stress impaired syndrome known as WPS (Fabiansson, 1986, Andersson, 1988)

WPS may have economic importance in many large commercial pig farms. Usually, it occurs 2 weeks post-weaning, and up to 7% of all weaners may exhibit a decrease in weight gain of more than 50% when compared with normal piglets of the same origin and age while maintained in the same environment. The typical clinical signs of pigs suffering from the WPS are a much reduced weight gain, prominent dorsal spine and increased hair coat. Haematological studies have shown a lowered plasma activity of alkaline phosphatase and a decreased plasma concentration of zinc, frequently combined with a non-specific eczema of the back of the wasting pig (Martinsson et. al., 1976). The hyperactivity of the pituitary-adrenocortical axis was indicated by the observation of hypertrophied adrenal glands (Martinsson et. al., 1978). The increased plasma concentration of non-bound cortisol in wasting pigs (Albinsson and Andersson, 1988) indicates that an anti-anabolic state may be present in these pigs, i.e. amino acids are directed into the gluconeogenetic pathways at the expense of normal protein metabolism. This means that ingested feed is preferably used for energy production instead of protein synthesis, thus making feed less available for body growth (Kyriakis and Andersson, 1989). Furthermore, the finding of thymus gland atrophy (Martinsson et. al., 1978) may be indicative of a stress-induced deterioration in immune function (Hara et. al., 1981). Therefore, resistance to diseases is diminished, mortality is high and poor growth/feed efficiency follows the pig up to the age of slaughtering (Fabiansson, 1986).

Before analyzing aspects of the prevention and treatment of the WPS, it is important to mention two other stress induced diseases of the early weaned piglet: Non Infectious Diarrhoea and the Post Weaning Diarrhoea Syndrome. Weaning of piglets at the age of 3-4 weeks, often leads to the development of non infectious diarrhoea, followed by the malabsorption syndrome (English et. al., 1978, Kyriakis, 1984). The malabsorption syndrome is caused by the stress of weaning and the simultaneous change of diet. Normally, this results in the following digestive alterations, to a greater or a lesser degree: 1) an increased excretion of fatty acids in the faeces, 2) an increased output of carbohydrates in the faeces, 3) watery faeces, 4) degenerative changes in the villus structure of the small intestine such as reduced number of enterocytes and decline of brush border enzymes (English et. al., 1978, Kyriakis, 1981). In most cases, opportunistic pathogens take advantage of the presence of non-infectious diarrhoea and the malabsorption syndrome. The most common pathogens which cause post weaning diarrhoea syndrome are enterotoxigenic strains of *E. coli* and *Rotavirus,* and as result of this syndrome there is a high fluid /electrolyte secretion from the small intestinal mucosa into the lumen of the intestine, severe diarrhoea, biochemical abnormalities and dehydration.

As already mentioned, a number of different social and environmental stressors may contribute to the development of WPS in early weaned pigs. In order to reduce these stressors to the minimum, efforts must be made regarding the areas of housing, nutrition and hygiene (Kyriakis, 1989). The housing requirements for the early weaned pig are already well known and established (Brent, 1985). The final composition of the diet after weaning must be based upon current knowledge of physiopathology of the digestive system of the piglet in addition to nutritional requirements during the same period (English et. al., 1978, Whittemore, 1999). The health status of the weaned piglet is also related to the hygiene standards which are set in the weaning house starting from the 'all in all out' husbandry system, proper disinfection techniques, moving and regrouping piglets in the correct way, using low densities of animals to reduce the rate of diseases and controlling the number of visits to the minimum (Taylor, 1999). Furthermore, other prevention and therapeutic measures includes (1) vaccination against specific infectious diseases which help the early weaned piglets to overcome health problems, (2) the use of antimicrobial agents applied, either therapeutically, prophylactically or metaphylactically and (3) fluid electrolyte supporting therapy based upon the use of glucose, sodium bicarbonate and vitamin C added to the drinking water (Kyriakis, 1983, 1984, Taylor, 1999).

One interesting feature regarding the treatment and the prevention of WPS is that the syndrome is frequently successfully treated with amperozide (Kyriakis and Andersson, 1989). Amperozide (chemical name: 4-[4,4-bis(p-fluorophenyl)butyl]-N-ethyl-1-piperazinecarboxamide) is a typical psychotropic drug with specific effects on nerve transmission in the limbic part of the brain (Pettersson et. al., 1987). Neurochemical studies have shown that amperozide has very weak affinity for the postsynaptic dopamine receptors in striatum but is a potent antagonist on 5HT2 receptors in the limbic part of the brain (Svartengren and Christensson, 1985). Behavioural studies have shown effects on dopaminergic and serotonergic nerve transmission in the limbic system. Furthermore, amperozide has potent anxiolytic properties as evident from its effects on emotion behaviour in a number of animal conflict models. The most prominent effect of amperozide however, indicating the limbic profile of action, is its effect in various experimental models of aggression. The total lack of interference with motoric skills and the absence of ataxia, underline the selective effect of amperozide on aggressive behaviour. The effect of amperozide markedly differs from that of azaperone and acepromazine whose effects on aggression are due to unspecific sedation and motoric impairment (Christensson and Gustafsson, 1985, Andersson et. al., 1987, Björk et. al., 1988). Finally, amperozide seems to interact with central neuroendocrine processes involved in the regulation of stress reactions (Andersson and Albinsson 1985).

Kyriakis and Andersson (1989) studied the effect of amperozide in an animal model exhibiting hyperactivity in the pituitary-adrenocortical axis, as is the case of WPS. The study was carried out at the largest commercial pig farm in Greece with around 3000 sows in production. A total of 100 homebred hybrid piglets at the age of 25-30 days, suffering from typical WPS were used in the study. The pigs were weaned at 17-20 days by moving them from the farrowing pens and mixing them in the flat decks on the farm. At the time of weaning all piglets in the trial were clinically healthy. Half of the pigs were treated orally with amperozide (2,5 mg/Kg body weight) and the control group was treated with long-acting oxytetracycline and vitamins (A+D3+E). There were significant improvements in average daily weight gain ($P < 0.05$) and feed conversion ratio ($P < 0.01$) in amperozide treated pigs compared to controls. Furthermore, there was a significant reduction ($P < 0.05$) in mortality in amperozide treated groups. Within 2 weeks following amperozide treatment the clinical signs of the WPS had disappeared and the pigs grew at the same rate as the healthy ones (Kyriakis and Andersson, 1989).

THIN SOW SYNDROME (TSS)

The thin sow syndrome (TSS), first reported over 30 years ago (MacLean, 1968) occurs particularly in stalled sows of lean breeds, in which weight loss occurs in lactation and early pregnancy, followed by failure to regain weight. This may become progressive and result in poor fertility and eventually in emaciation and death. The syndrome is also not uncommon in loose housed sows where individual feeding is not practised (Taylor, 1999). In Denmark, it was found that about 30% of sows in one abattoir were ‘thin sows’ (Kristensen, 1982).

The aetiology may be related to a combination of causes such as parasitism, low environmental changes, inadequate and/or inappropriate feed intake during lactation (Taylor, 1999). Nevertheless, in many field cases when all of the above factors are not present, the thin sow syndrome still occurs causing economic problems in farms using intensive methods of production. Thus, the syndrome may have an aetiology related mainly to social and environmental stressors during the very critical period of lactation, especially in the cases of large litters (Kyriakis et. al., 1990).

In the terms of behaviour and welfare, stress is often perceived as the consequence of the restraint of animals which occurs in intensive production units. Particular restraints are seen with tethered and stall-housed pregnant sows which cannot turn around, and have limited (less than 1m) forward and backward movement, and also with lactating sows similarly restrained in farrowing crates. On the other hand, increased movement freedom levels may themselves bring stress, particularly when adult sows are placed into groups. Adult sows may fight vigorously when placed into novel social groups, as is often the case after weaning. As such fighting may also occur around the time of oestrus, and during the first 3 weeks of the pregnancy, losses of embryos can result in the reduction of litter size. Sow groups are often formed with one dominant sow, which eats greedily and becomes large and fat. There is also likely to be one or two subservient sows which actively avoid competitive feeding situations and, in addition to being wounded and bruised, may become emaciated (Whittemore, 1998). The underling mechanisms of stress at the individual animal level are similar to those previously described (wasting pig syndrome).

All phases of the reproductive cycle of a sow are interrelated and therefore the feeding programme in one phase can have significant effects on the sow's performance and appetite in another phase. It has been shown that feeding levels during gestation that promote backfat levels at parturition of 20mm P2 (P2 backfat is measured at the level of the last rib 65mm from the mid-line of the sow) or more, may result in reduced sow feed intake during lactation, especially in early lactation. High level of feed intake during gestation will either decrease insulin secretion during lactation or increase insulin resistance. This may reduce peripheral glucose utilization and reduce mobilization of fat stores, which in turn will reduce feed intake. The resulting reduced feed intake during lactation will lead to an increased weight and condition loss of the sow (Ahern, 1999).

The clinical signs of the syndrome include emaciation in 30-90% of sows and boars in a herd, associated with hypothermia (35.5-38 °C), depraved appetite, restlessness, apathy and later on difficult in rising. The skin may be dirty and greasy and there may be surface abrasions. As the condition progresses, failure to return to oestrus and permanent infertility may occur (Taylor, 1999). This condition also includes features such as overactivity, and consumption of non-nutritive substances, for example straw (Owen, 1995). Analysis of blood and faeces from 'thin sows' did not show any significant changes, compared to normal sows. It was concluded that the clinical and the haematological changes had similarities to the situation of hyperfunction of the thyroid gland (Kristensen, 1982). Also, in many clinical cases of this syndrome, renal vascular lesions, unrelated to parasitism, have been identified and this is believed to be a stress induced situation (MacLean, 1968). There is much present interest in the possibilities for systems of production which increase potential movement and space whilst avoiding some of the new stressors which may be associated with it. Such stressors will usually include: (1) avoidance of injury and disease, (2) generous provision of feed and water, (3) small group size, (4) solid flooring for at least a part of the pen, (4) use of bedding (straw), (5) low density stocking, (6) individual feeding for breeding sows and stability of animal groups by obtaining constancy of individuals in the group (Whittemore, 1998). Furthermore, environmental temperatures should be restored to normal and anthelminthic treatment should be considered (Taylor, 1999).

It is of great importance to ensure that adequate quantities of feed are given to all affected animals (Taylor, 1999). It has been suggested that gilts and sows should be bred with 16 to 17 mm P2 backfat and should remain with roughly that level of P2 backfat at successive weanings. This could be achieved by having sows of all parities with about 19mm backfat at farrowing. These sows could then lose 3mm backfat during lactation and still have 16mm backfat at weaning. The nutrient and energy requirements of a gestating sow will depend on her weight, backfat level and the amount of weight gain she will need during gestation to meet a target level of 19mm P2 at farrowing. From the data available (NRC, 1998, Whittemore, 1998) reasonable estimates of the energy, protein and lysine requirements of gestating sows can be calculated (Ahern, 1999). Finally, lactation feed requires to be provided at adequate levels. Lactating sows need to eat as much as they can, but in many cases even that is not enough (Whittemore, 1999).

As we have already mentioned TSS has an aetiology which is related mainly to social and environmental stressors during the very critical period of lactation, especially in cases of large litters. Therefore, the syndrome may be related to the wasting pig syndrome (WPS) another stress induced disease (Kyriakis et. al., 1990). The observation that in the case of WPS amperozide was used with very good results (Kyriakis and Andersson, 1989) lead Kyriakis et. al. (1990) to investigate the effect of amperozide on the weight gain and health status of sows with TSS. The field study was conducted in a commercial pig farm in Greece, with around 3000 sows in production. The prevalence of the thin sow syndrome is approximately 6% of the sows at post-lactation. The farm follows the genetic programme of a very well known, European hybrid brand name of lean breed. Sixty sows, suffering from typical post-weaning thin sow syndrome, were divided into three groups: (1) 20 as negative controls (NC), (2) 20 as positive controls (PC) treated with vitamins, trade elements and antibiotics and (3) 20 injected with amperozide (2mg/kg body weight). Amperozide treated sows fully recovered ($P< 0,05$) and 80% became pregnant, while figures for the PC and NC groups were only 15% and 10% respectively. Mortality was up to 50% in the NC, 40% in the PC and only 15% in the amperozide treated group ($P<0.05$) (Kyriakis et. al. 1990).

Additionally, Kyriakis (1991) conducted another trial to determine the effect of amperozide on sow-litter productivity while it was used in sows at farrowing and, or, at weaning. The study was conducted in a commercial pig farm in Greece with 400 sows under production. The breed of sows was F1: Landrace cross Large White and the sows were served by Belgian Landrace boars. A total of 64 sows divided into four groups: (1) 16 as untreated controls, (2) 16 treated with a single dose of amperozide at farrowing, (3) 16 treated with a single dose of amperozide at weaning and (4) 16 treated with two single doses of amperozide at both farrowing and weaning. No clinical outbreak of mastitis, metritis, agalactiae complex was recorded in the amperozide groups dosed at farrowing. The number of "empty" days was decreased (about three days) in sows treated with amperozide at weaning. Furthermore, pre-weaning mortality decreased (8 %) and piglet growth rate improved when sows were treated with amperozide at farrowing (Kyriakis, 1991).

The results from both studies suggest that amperozide treats TSS and improves the health status and productivity of sows by reducing their emotional responses to novel or threatening situations (Kyriakis et. al., 1990, Kyriakis, 1991).

ANIMAL BEHAVIOUR AND ANOREXIA NERVOSA – CONCLUSIONS

Eating disorders are currently increasing alarmingly in the Western developed nations (Goodrick et. al., 1996). It is commonplace to refer to the epidemic of eating disorders in Western women but few are aware of the parallels on the farm and especially at the modern pig intensive production (Treasure and Owen, 1997). As we have already described, pigs, in particular those which have been bred for extreme leanness, can develop irreversible self starvation and emaciation (MacLean, 1968, Kyriakis and Andersson, 1989, Kyriakis et al., 1990). Anorexia in pigs develops after weaning as the wasting pig syndrome (WPS) or after farrowing and/or weaning as the thin sow syndrome (TSS). WPS and TSS are related mainly to social and environmental stressors during the very critical period of lactation and weaning and are widespread under some modern intensively pig husbandry systems. In the largest pig farm in Greece with around 3000 sows in production, stocked with a very well known European hybrid brand name of lean breed, the prevalence of the condition is 6% (Kyriakis et al. 1990).

The clinical signs observed in both the two syndromes show an uncanny resemblance to those of anorexia nervosa. In particular affected animals restrict their normal feed intake, although some consume large amounts of straw. The affected animals spend more time on non-nutritive hyperactive behaviour and they are easily recognized as they develop a prominent dorsal spine and their hair becomes coarse and long and many of them do not return to heat. Few abnormalities are found at post-mortem examination and these include enlargement of adrenal glands and thymus atrophy. The early separation of sow and piglets and mixing of unacquainted pigs appear to be the trigger. When pigs from different litters are placed together for the first time they start fighting and over the next 24 to 48 hours a new dominant hierarchy is established. MacLean (1968) in his detailed study reported the possibility of a hereditary relationship because certain family lines were more severely affected than others (MacLean 1968).

Having described WPS and TSS the outcome question is whether WPS and TSS could serve as an animal model for anorexia nervosa. There have been recent proposals to simplify the diagnostic criteria of anorexia nervosa, as it is recognized that the features of the psychopathology vary across time and cultures. A proposed definition included the following : "subjects who become emaciated through restricting their dietary intake for whatever reason, this restriction is deliberate, and the resulting state is positively valued by the subject" (Szmuckler and Patton, 1994). Another proposed description says: "the patient avoids food and induces weight loss by virtue of a range of psychosocial conflicts whose resolution she perceives to be within her reach through the achievements of thinness and/or the avoidance of fatness" (Russell 1994). Omitting the subjective attribution of motivation, it could be argued that wasting pigs and thin sows fulfill these broader criteria (Treasure and Owen, 1997).

Accepting this proposition it can be said that there is an intriguing possibility that the condition has an analogous genetic basis in both species and that the condition has only emerged in pigs since the relatively recent intensive selective breeding for leanness (Treasure and Owen, 1997). Owen (1990, 1992) has argued, drawing evidence from farm animals, that anorexia nervosa may rise primarily as an extreme variant of body composition. Farmers selectively breed their stock to optimize carcass composition. In recent years, particularly in pigs, consumer aversion to fat has meant that leanness has been favoured over fatness. Such selective breeding has led to the uncovering of recessive traits that produce extremes like the halothane gene. The halothane gene is of interest because it is associated with leanness, sensitivity to stress and porcine stress syndrome and the condition is seen in both pigs and humans. This gene or genes nearby sensitize the animal to stress, and sometimes affect appetite and reproduction (Treasure and Owen, 1997). Finally, as we have already mentioned, body composition is reviewed as a composite of several traits each with their distinctive genetic basis, including major effects of genes at single loci (Owen, 1999).

One interesting feature regarding the treatment and/or the prevention of WPS and TSS, is that the syndrome is frequently successfully treated with amperozide (Kyriakis and Andersson, 1989, Kyriakis et. al., 1990, Kyriakis, 1991). It is not surprising that a drug that acts on central serotonin (5HT) has been found to be effective, suggestive the pivotal role of serotonin (in particular 5HT2 receptors) in central control of appetite. 5HT, is also implicated in the central control of locomotor activity and sexual behaviour, both of which are components of the clinical syndrome of anorexia nervosa. Therefore, abnormalities in central 5HT may result in a vulnerability to develop anorexia nervosa. Unfortunately, pharmacological approaches to treatment of anorexia nervosa have failed to impress clinicians with their benefits. However it remains likely that psychotherapy will be essential to understand the difficulties that produce the stress response in order to help the patient develop more effective coping strategies and to combat the pervasive value placed on thinness in our society. It is ironic that pigs, which are not renowned for their anorexic eating attitudes, may stimulate several researches leading to determination of the etiology of anorexia nervosa (Treasure and Owen, 1997).

Acknowledgements

The author acknowledges Mr.Ch.C.Miliotis BVM, post-graduate student at the Clinic of Productive Animal Medicine of the Faculty of Veterinary Medicine at the Aristotelian University of Thessaloniki, for his help during the writing of this chapter.

REFERENCES

Ahern F. Feeding the Gestating Sow. *International PIGLETTER* 1999; 19: 8.

Albinsson A, Andersson G. Subclinical characteristics of the wasting pig syndrome 1988 (Submitted)

Algers B, Jensen P, Steinwall L. Effects of teat quality on weight and behaviour changes in pigs at weaning and regrouping. *Summaries of the International stress symposium in Malmoe.* May 31st; Malmoe. 1988.

Andersson G, Albinsson A, Olsson NG. Amperozide inhibits aggressive behaviour and the development of stress related diseases. *Proceedings of the 9th International Pig Veterinary Society Congress, Barcelona.* 1986.

Andersson G, Olsson N-G, Albinsson A. The effects of amperozide on social stress in pigs. *XXIII World Veterinary Congress,* Montreal. 1987.

Andersson G. The effects of amperozide in social stress in pigs. *Summaries of the International stress symposium in Malmoe.* May 31st; Malmoe. 1988.

Andersson L, Haley CS, Ellegren H, Knott SA, Johansson M, Andersson K, Andersson-Eckllund L, Edfors-Lilja I, Fredholm M, Hansson I, Hakansson J, Lundstrom K. Genetic mapping of quantitative trait loci for growth and fatness in pigs. *Science* 1994; 263: 1771-1774.

Andersson LB. Genes and Obesity. *Anim Med* 1996; 28: 5-7.

Björk A, Olsson N-G, Christensson E, Martinsson K, Olsson O. Effects of Amperozide on biting behaviour and performance in restricted-fed pigs following regrouping. *J Anim Sci* 1988; 66: 669-675.

Black JL. "Modeling energy metabolism in the pigs – critical evaluation of a simple reference model" In *Modeling Growth in the Pig.* Moughan, PJ, Verstegen, MWA, Visser-Reyneveld, MI, eds. The Wageningen, Netherlands: Wageningen Pers. 1995.

Brambell RWP. Report of the technical committee to enquire into the welfare of animals kept under intensive livestock husbandry systems. *Command Paper* No.2.836. London: H.MS.O. 1965.

Brent G. *Housing of Pigs.* Ipswich UK: Farming Press. 1986.

Chapple RP. "Effect of stocking arrangement on pig performance". In *Manipulating Pig Production IV.* Victoria: Australian Pig Science Association. 1993.

Christensson E, Gustafsson B. Amperozide, a novel psychotropic compound with specific effect on limbic brain areas. *Acta Vet Scand* 1985; 124 (suppl 542): 281.

De Haer LCM, Merks JWM. Patterns of daily food intake in growing pigs. *Anim Prod* 1992; 54: 95-104.

English PR, Smith WJ, MacLean A. *The sow: improving her efficiency.* Ipswich, UK: Farming Press. 1978.

Everby GS, Rosenfeld R. "What is Stress?" In *The Nature and Treatment of the Stress Response.* New York: Plenum Press. 1981.

Fabiansson S. The SIG-PIG syndrome and its cause – social pressures at regrouping. *Proceedings of the 9th International Pig Veterinary Society Congress,* Barcelona. 1986.

Fujji J, Otsu K, Zorato F, De Leon S, Khanna VK, Weilre JE, O'Brien PJ, MacLennan DH. Identification of a mutation in porcine ryanodine receptor associated with malignant hyperthermia. *Science* 1991; 253: 448-451.

Goodrick GK, Poston WSC, Foreyt JP. Methods for voluntary weight-loss and control-update. *Nutrition* 1996; 12: 672-676.

Grimble RF. Nutrition and cytokine action. *Nutr Res Revs* 1990; 3: 193-210.

Gull WW. Anorexia nervosa (apepsia hysterica, anorexia hysterica). *Transactions of The Clinical Society of London* 1874; 7: 22-28.

Hara C, Manabe K, Ogawa N. Influence of activity-stress on thymus, spleen and adrenal weights of rats: Possibility for an immunodeficiency model. *Physiol Behav* 1981; 27: 243-248.

Hefner DL. Nutritional management of the early weaned piglet. *Proc Am Assoc Swine Practitioners,* Indianapolis. 1987.

Herny JP, Stephens PM. The social environment and essential hypertension in mice: possible role of innervation of the adrenal cortex. *Prog Brain Res* 1987; 47: 263.

Johnson RW, von Borell E. Lipopolysaccharide-induced sickness behaviour in pigs is inhibited by pretreatment with indomethacin. *J Anim Sci* 1994; 72: 309-314.

Kristensen SP. Den magre sos syndrome. Praeliminaer undersØgelse af blodparametre for tidig diagnostik af afmagrede sØer. *Danske Veterinar Tidskrift* 1982; 65: 73-79.

Kyriakis SC. *Investigations of the Post Weaning Diarrhoea Syndrome*. Post Doctoral Thesis. University of Thessaloniki, 1981.

Kyriakis SC. Post weaning diarrhoea syndrome (PWDS) of piglets. A new therapeutic approach with supporting therapy. *Pig News and Information* 1984; 4: 23-27.

Kyriakis SC, Andersson G. Wasting pig syndrome (WPS) in weaners – treatment with amperozide. *J Vet Pharmacol Ther* 1989; 12: 232-236.

Kyriakis SC. New aspects of the prevention and/or the treatment of the major stress induced diseases of the early weaned pigs. *Pig News and Information* 1989; 10: 177-181.

Kyriakis SC, Martinsson K, Olsson N-G, Björk A. Thin sow syndrome (TSS): The effect of amperozide. *Br Vet J* 1990; 146: 463-467

Kyriakis SC. Observations on the action of amperozide: are there social influences on sow-litter productivity ? *Res Vet Sci* 1991; 51: 169-173.

Kyriakis SC, Alexopoulos C, Giannakopoulos K, Tsinas AC, Saoulidis K, Kritas SK, Tsilogiannis V. Effect of a paramunity inducer on reproductive performance of gilts. *J Vet Med* 1996; 43: 483-487.

Kyriakis SC, Tzika ED, Lyras DN, Tsinas AC, Saoulidis K. Effect of an inactivated Parapoxvirus based immunomodulator (Baypamun) on post weaning diarrhoea syndrome and wasting pig syndrome of piglets. *Res Vet Sci* 1998; 64: 187-190.

Martinsson K, Ekman L, Jönsson L. Hematological and biochemical analyses of blood and serum in pigs with regional ileitis with special reference to pathogenesis. *Acta Vet Scand* 1976; 17:233-243.

Martinsson K, Ekman L, Löfstedt M, Figueiras H, Jönsson L. Organ weights and concentration of zink in different tissues of wasting pigs and pigs with regional ileitis. *Zentralblaltt Für Veterinärmedizin* A 1978; 25: 570-578.

McClone JJ, Salak JL, Lumpkin EA, Nicholson RI, Gibson M, Norman RL. Shipping stress and social status effects on pig performance, plasma cortisol, natular killer cell activity, and leukocyte numbers. *J Anim Sci* 1993; 71: 888-896.

McGlone JJ, Désaultés C, Morméde P, Heup M. "Genetics of Behaviour" In *The Genetics of the Pig*. Rothschild, M. and Ruvinsky, An. eds. New York: CAB International. 1997.

MacLean CW. The thin sow syndrome. *Vet Rec* 1968; 83: 308-316.

McRae JC, Lobley GE. Physiological and metabolic implications of conventional and novel methods for the manipulation of growth and production. *Livestock Prod Sci* 1991; 27: 43-59.

Mitichashvili RS, Tikhonov VN, Bobovich VE. Immunogenetic characteristics of the gene pool Kakhetian pigs and dynamics of population parameters during several generations. *Sov Genet* 1991; 26: 855-863.

Morgan CA, Nielsen BL, Lawrence AB, Mendl MT. "Describing the Social Environment and its Effects on Food Intake and Growth" In *A Quantitative Biology of the Pig*. Kyriazakis, I. ed. New York: CABI Publishing. 1999.

National Research Council. *Nutrient Requirements of Swine 10th ed.* National Academy Press. Washington DC, USA. 1998.

Oliverio A. "Endocrine aspects of stress: central and peripheral mechanisms". In *Biology of Stress in Farm Animals: An Integrative Approach*. Wiepkema, PR. and van Adrichem, PWM. Dordrecht: Martinus Nijhoff Publishers. 1987.

Owen JB. Weight control and appetite: nature over nurture. *Animal Breeding Abstracts* 1990; 58: 583-591.

Owen JB. Genetic aspects of appetite and feed choice in animals. *J Agric Sci* 1992; 119: 151-155.

Owen JB. " Models of Eating Disturbances in Animals" In *Neurobiology in the Treatment of Eating Disorders*. Hoek, HW, Treasure, JL, Katzman, MA, eds : John Wiley & Sons Ltd. 1998.

Owen JB. Genetic Aspects of Body Composition. *Nutrition* 1999; 15: 609-613.

Pettersson G, Gustafsson B, Svartengren J, Cristensson E. Effects of amperozide a novel psychotropic agent, on dopaminergic neurotransmission in brain. *6th International Catecholamine symposium*, Jerusalem. 1987.

Pond WG, Maner JH. *Swine Production and Nutrition* Westport, Connecticut: Avi Publishing Company, Inc. 1984.

Ratner SC, Boice R. "Effects of domestication on behaviour" In *The Behaviour of Domestic Animals* 3rd edn. Havez, ESE. Ed. London: Baillière Tindall & Cassel. 1976.

Rothschild M, Ruvinsky AM. *The Genetics of the Pig*. New York: CAB International. 1997.

Russell GMF. Anorexia nervosa through time. In *HandBook of eating disorders: Theory, treatment, and research*. Szmuckler, GI, Dare, C, Treasure, JL. eds. Chichester: Wiley. 1994.

Svartengren J. Christensson EG. Chronic amperozide treatment regulates 5-HT2 receptors labelled by ^{3}H-Ketanserin. *Acta Vet Scand* 1985; 124 (suppl 542): 221.

Szmuckler GI, Patton G. "Sociocultural models of eating disorders". In *HandBook of eating disorders: Theory, treatment, and research*. Szmuckler, GI, Dare, C, Treasure, JL. eds. Chichester: Wiley. 1994.

Taylor DJ. *Pig Diseases*, 7th edn. Suffolk: St Edmundsbury Press Ltd. 1999

Treasure JL, Owen JB. Intriguing Links Between Animal Behaviour and Anorexia Nervosa. *Int J Eat Disord* 1997; 21: 307-311.

Varley M, Stedman R. "Stress and Reproduction". In *Principles of Pig Science*. Cole, DJA, Wiseman, J, Varley, MA. eds. Nottingham: Nottingham University Press. 1994.

Vogeli P. *The behaviour of fattening performance parameters and quantitative and qualitative carcass characteristics of two divergent lines of pigs using index selection procedures*. Doctoral Thesis, Swiss Federal Institute of Technology, Zurich, Switzerland 1978.

Whittemore, C. *The Science and Practice of Pig Production*. Oxford: Blackwell Sciences. 1998.

Chapter 13

LABORATORY ANIMAL MODELS FOR INVESTIGATING THE MECHANISMS AND FUNCTION OF PARASITE-INDUCED ANOREXIA

Julian G. Mercer[1] and Leslie H. Chappell[2]

Aberdeen Centre for Energy Regulation and Obesity (ACERO), [1]Rowett Research Institute, Bucksburn, Aberdeen, AB21 9SB, United Kingdom, and [2]Department of Zoology, University of Aberdeen, Aberdeen AB24 2TZ, United Kingdom.

ABSTRACT Reduced voluntary food intake, or anorexia, is commonly observed in mammalian hosts infected with parasitic worms. Despite the frequent occurrence of anorexia, the functional significance of reduced food intake is still uncertain. It seems likely that, acutely, anorexia benefits the host, whereas more chronic malnutrition is detrimental. The mechanisms underlying this voluntary excursion into negative energy balance have not been extensively studied. We have investigated the endocrine and neuroendocrine consequences of infection of the laboratory rat with the hookworm-like nematode, *Nippostrongylus brasiliensis*. This parasite induces a biphasic anorexia in its host. Although elevated levels of the adipocyte hormone, leptin, may be causally implicated in infection-induced anorexia, most of the perturbations in the concentrations of leptin, insulin and corticosterone in the bloodstream, and in expression of the anabolic neuropeptide, neuropeptide Y, in the hypothalamus, appear to be secondary to the state of negative energy balance, and to reflect that state. Other factors, such as cytokines, released during the infection may block the normal response to elevated feeding drive encoded by peripheral feedback and CNS integratory systems. Further investigation of the ways in which parasites manipulate their hosts, and of host responses, will allow intervention strategies to be more appropriately evaluated.

INTRODUCTION

On a global scale parasite-induced anorexia probably represents one of the most common as well as the most economically and medically costly disorders of eating behaviour. However, the real magnitude of this

J.B. Owen et al. (eds.), Animal Models – Disorders of Eating Behaviour and Body Composition, 223–241.

cost in terms of healthcare resource and livestock production is difficult to assess accurately. Despite the wide range of parasites of medical and veterinary importance known to induce anorexia (reduced food intake) in their host, little is understood of the physiological mechanisms that underlie this phenomenon. Moreover, it is not clear whether these mechanisms are the same in all host-parasite interactions, or whether parasite-induced anorexia has shared elements with anorexias induced by bacterial and viral infections or tumours. Additionally, and most importantly, what precisely is the functional significance of this behavioural modification? The drive to consume food is one of the fundamental motivations in mammalian life. Many mammals have only limited energy reserves, stored in the form of adipose tissue. Small mammals, for example, can only survive a few days in the complete absence of food and while an animal may go out each day in search of food, it is not inevitable that any individual will be successful. Thus for an animal to voluntarily forgo consumption of food that is freely available will of necessity require very powerful counteracting signals. The question that forms the basis of this review of recent research employing laboratory model systems is "By what mechanisms do parasitic infections reduce the appetite of the host?".

In attempting to investigate these issues in medical and veterinary contexts, we are faced with differences in physiology and metabolism between ruminant and monogastric mammals. The major part of the work discussed herein has been performed in a laboratory rodent and with a parasite species that acts as a model for species of clinical and veterinary importance. Although much of the work on the metabolic and nutritional consequences of parasitic infection has involved ruminant species in which large body size is a positive attribute, other experimental parameters are probably better studied in a small rodent. This enables the experimenter to draw upon the accumulated body of knowledge in topics as diverse as neuroanatomy, energy balance regulation and immune function that are well developed in laboratory rodents. Many regulatory mechanisms, and particularly those relating to food intake and body weight regulation are well conserved within mammals, and studies should employ the host species that is most appropriate for the specific question being addressed.

Nutritional and metabolic consequences of infection with helminth parasites

The phenomenon of parasite-induced anorexia has been best described for helminth infections. The helminths includes worm parasites belonging to three invertebrate phyla, Platyhelminthes, Nematoda and Acanthocephala. Infections with these worms are widespread in mammals in general, ranging from wild and domesticated animals to man, where infections of individual parasite species may occur in hundreds or thousands of millions of people. These infections, which may survive in the host for a few days or for many months, have both economic and health impacts, affecting farmed ruminants across the globe and people who inhabit the tropical and subtropical third world. Both clinical and sub-clinical infections can have a negative impact on animal production. This impact may be directly at the level of food intake, where anorexia can account for a large proportion of the weight gain differential between parasitised and uninfected ruminants. Additionally, altered metabolism and nutritional demands also contribute to production deficits, including, where it occurs, the energetic costs of fever. For example, intestinal infections may be characterized by elevated protein metabolism in the gastrointestinal tract, thereby diverting amino acid resources from other body organs. The amplification of digestive secretions and mucosal protein synthesis coupled with tissue damage and digestive/absorptive dysfunction may cause nitrogen loss to the environment and other post-ingestive consequences. Many parasites of man are also recognized to contribute to cycles of malnutrition and infection through loss of appetite and altered gut physiology, the latter leading to malabsorption and nutrient loss.

Research into the pathology of infection with nematode species in domestic ruminants has recently focused on food intake depression, gastrointestinal function and protein metabolism. These different facets of the pathology may be directly related. For example, appetite suppression in cattle infected with the nematode *Ostertagia* is accompanied by elevated blood concentrations of the hormone, gastrin, and associated changes in gut motility (Fox, 1997). However, elevated gastrin, which in this example is postulated to act as a peripheral satiety factor, is not a ubiquitous occurrence in intestinal nematode infections. Cholecystokinin (CCK) is another peripheral peptide that reduces food intake when injected systemically. Whereas the evidence supporting the involvement of gastrin in anorexia in the specific case of *Ostertagia* infection is strong, earlier studies suggesting that CCK might be elevated during parasite-induced anorexia have not been substantiated (Coop & Holmes 1996). However, when considering the signals that control food intake and energy homeostasis, the complexity, redundancy and hierarchy of the system must be taken into account. Attention must also be directed towards the central nervous system, and in particular on those brain centres that integrate peripheral signals of neural and hormonal origin.

Host defence mechanism or parasite advantage?

In essence the question here boils down to "Who benefits?" from parasite-induced anorexia. On casual inspection the need to sustain the nutritional and metabolic requirements of a considerable number and/or biomass of invasive agents of another species, and to mount an immune response, would be expected to increase the energy demands of the host mammal. Although there are a few specific examples of host-parasite interactions that are associated with weight gain in the vertebrate definitive host, for the most part effects of parasitic infection on energy balance are characterized by a relative anorexia and associated weight loss and/or growth restriction. This paradoxical situation immediately raises the issue of which of the 'partners' in the host-parasite relationship drives the response and thereby benefits from the negative energy balance of the host. According to Langhans (2000) an anorexic response to infection may initially be of benefit to the host while longer term self-imposed food restriction must harm the host.

While ethical, clinical and humanitarian considerations constrain investigations of worm-induced anorexia in man, large animal studies and laboratory models are disclosing the nature and significance of the phenomenon. As more information is accumulated on nutrition, infection and immunity it may be possible to unravel the complex interactions that lead to the infected host consuming less food. A full knowledge of these proposed pathways might permit the function of this phenomenon to be determined or allow the development of interventions that can ameliorate the nutritional symptoms. Furthermore, the increasing prevalence of drug resistance in helminth parasites will hinder their direct removal by conventional chemotherapy and it may become necessary to develop alternative strategies to reduce the effects of parasitism. The wisdom of a strategy to overcome parasite-induced anorexia will depend upon the origin of this anorexia and its functional significance (Kyriazakis, Tolkamp & Hutchings, 1998). Whether anorexia is a protective response or a deleterious pathological phenomenon is clearly a critical issue in the management of domestic livestock where parasitic infections are a constraint on animal production. Were the anorexia directly induced by the parasite, circumventing its expression could be to the advantage of the host both in terms of its overall plane of nutrition and in blocking whatever advantage might be gained by the parasite from the anorexic state of its host. Conversely, the consequences of intervening to overcome an anorexia that formed part of the host's defence against the parasite could be deleterious to the host and therefore unacceptable. There is some evidence to suggest that the infected animal may be compromised by attempts to overcome the anorexia early on in an infection, but that over longer periods of time relative malnutrition may delay recovery and increase susceptibility to further infection (Langhans, 2000).

A number of possible functional explanations of parasite-induced anorexia have been proposed (Kyriazakis, Tolkamp & Hutchings, 1998). If anorexia affords advantage to the parasite, acting as a strategy to run down host nutrition or to adapt its environment beneficially by, for example, impairing the host immune response, then we might predict that this would be a relatively ubiquitous, but timely, occurrence. Anorexia is indeed a consequence of a range of viral, bacterial and parasitic infections, the latter encompassing both protozoan and helminth species. The timing of the onset of anorexia is species-specific and appears to relate to the life cycle of the parasite. Where the parasite causes severe depression of food intake, and clinical disease is observed, negative energy balance of considerable magnitude would be expected to have a negative impact upon immune function. It is also possible to construct an argument that excesses of some nutrients may actually suppress immune responses, and that moderate anorexia might therefore be of some benefit (Kyriazakis, Tolkamp & Hutchings, 1998).

A plausible explanation for infection-induced anorexia in generality as well as in the specific context of parasitism is that anorexia is a part of the acute phase immune response of the host (Exton, 1997; Plata-Salaman, 1995; Langhans 2000). Here anorexia can be placed in an array of physiological modifications that prevent the development or reproduction of the infectious agent, or reduce the likelihood of detrimental metabolic effects of the acute phase response. The 'starvation' of the invading organism is one possible outcome here. A further possibility is that a relative anorexia could allow the host to limit further acquisition of parasites by reducing exposure to infectious agents acquired during feeding bouts, although it is difficult to see how this would relate to those infections acquired other than during feeding. Reduced food intake could facilitate more careful diet selection to maximise the ingestion of foodstuffs that enable the host to cope better with the metabolic consequences of parasitism, or maximise ingestion of beneficial plant compounds. Many parasites appear to evade the immune responses of their mammalian hosts through the manipulation of that response, and this undoubtedly contributes to the chronic nature and longevity of many worm infections. It is conceivable that the interaction between the infecting agent and its host at the level of nutrition and immunity could be quite different in infections of different duration that are either clinical or sub-clinical. Before the function of anorexia can be resolved, it will be necessary to disentangle the loop of parasite-induced increases in metabolic and nutritional demand, immune system activation, and feedback of food intake onto physiological systems, including the immune system.

Laboratory models of parasite-induced anorexia

The principal motivation for studying parasite-induced anorexia in the laboratory is to apply the knowledge gained of the fundamental underlying mechanisms to parasites of medical or veterinary importance. Most natural infections with parasitic helminths are likely to be acquired gradually or continually over periods of weeks or months, or on a seasonal cycle, as animals feed in an environment that is contaminated with infective agents. A number of different parasite species may be acquired simultaneously. This type of longitudinal infection period has been mimicked in ruminant species through daily 'trickle' infections of a constant dose of infective organisms. Such complex processes are less easy to model in the laboratory, and the available rodent models of parasite-induced anorexia do not readily lend themselves to this type of infection regime. As a consequence most laboratory infections consist of the introduction of a single large dose of parasite larvae. While this circumvents the problems of working with large agriculturally important species, it raises the issue of how applicable such studies will be to natural host-parasite relationships in the field. Although these models may have limitations, they nevertheless offer the best opportunity to investigate the causes and effects of anorexia at a mechanistic level and enable the experimenter to draw upon the growing knowledge of immune function and energy balance in small rodents.

Nippostrongylus brasiliensis in the laboratory rat

In order to study the mechanisms underlying parasite-induced anorexia we selected a host-parasite combination that was readily maintained in the laboratory, and which exhibited a robust, reproducible anorexia. The hookworm-like nematode *Nippostrongylus brasiliensis*, with the laboratory rat as its mammalian host, has a direct life cycle and causes an acute biphasic anorexia in rats that occurs despite the provision of food in excess (Fig. 1). The infective larvae (L3 stage) enter the host through the skin, are transported via the circulation to the heart and thence to the lungs (day 2 post-infection). Larvae are then coughed up and swallowed, entering the gastrointestinal tract. Upon maturation they begin egg production from day 5 post-infection. Nine days after infection the host initiates a very effective immune response that expels the parasites, such that by day 15 post-infection few worms are left in the intestine. The biphasic anorexia is linked to the life cycle of the parasite. The first, substantial, reduction in food intake takes place over the first 2 days post-infection when migrating parasites enter the lungs. Food intake then generally recovers, but not to normal levels, before the appearance of a second phase (days 6-8 post-infection). This second phase of anorexia is of variable severity and coincides with the establishment and maturation of adult worms in the small intestine. Following expulsion of worms, rats often display a compensatory hyperphagia and subsequently gain weight.

Many of our studies with this host-parasite model have employed the same basic protocol. Typically, rats were arranged into groups of three animals with similar body weights. Within each group, rats were then randomly assigned for either infection, pair-feeding or controls. Thus each infected rat had a weight-matched partner in both control and pair-fed groups. Rats in the infected group were injected with 45 L3 larvae/g body weight. The other two groups received an injection of tap water. Infected and uninfected control rats were allowed to feed *ad libitum* throughout, and food intake was recorded daily. Pair-fed rats were provided with a daily food ration equal to that consumed by their designated infected partner during the previous 24 hours. The anorexia induced by parasitic infection is by its nature a very heterogeneous phenomenon.

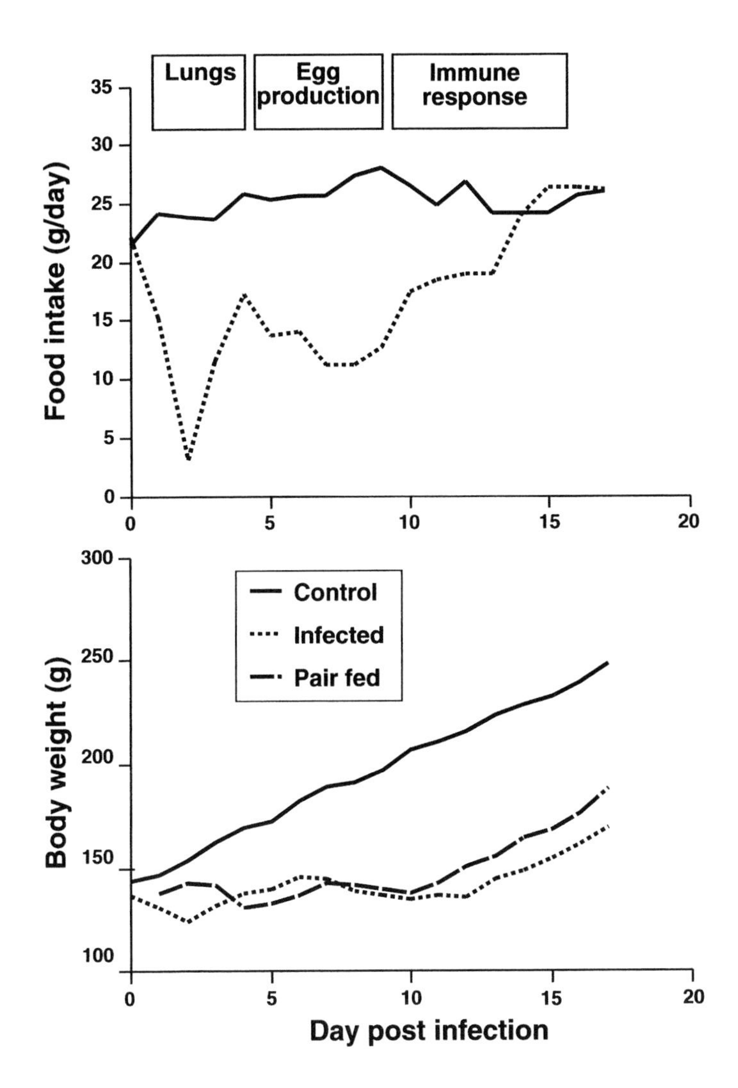

Figure 1. Typical trajectories of food intake (g per day) and body weight (g) in rats infected with *Nippostrongylus brasiliensis* on day 0 and their uninfected and pair-fed controls during the 17-day period following infection. Key features of the infection are indicated.

A theoretically identical inoculation of infective larvae inevitably gives rise to mature worm burdens of different numerical size, either as a consequence of variation in the size of the actual inoculum or in host susceptibility to infection. The effect on food intake of an infection is thus also variable. Consequently, in investigating the causes and effects of anorexia in the rat host we have treated each animal as an individual, with its own appropriate controls, rather than establishing pair-fed controls with food intake restricted to the average of the infected population.

Parasite-induced anorexia is a recognized consequence of many infections with parasitic helminths, but the biological determinants of this phenomenon are not well resolved. Laboratory models of parasite-induced anorexia which have been studied in some detail, such as *N. brasiliensis* (Ovington, 1985), have thus far given little insight into the mechanistic basis of this pathology, or indeed its functional significance. There are two fundamental issues to be addressed in the laboratory model of parasite-induced anorexia afforded by *N. brasiliensis* infection of the rat. Firstly are the mechanisms involved in the initial induction of anorexia, and secondly, given the power of the compensatory responses that are normally activated by negative energy balance, how this anorexia is maintained until the expulsion of the parasite. The biphasic nature of the anorexia observed in *N. brasiliensis* infections of the rat is suggestive of multifactorial mechanisms and different signalling pathways. This hypothesis is strengthened by the observation that standardised infections give rise to different patterns of Fos protein expression in the hindbrain at different points in the parasite life cycle (Castex *et al.*, 1998). Fos protein is the product of the immediate-early gene, c-*fos*, and is a valuable tool in identifying pathways of neuronal activation in response to defined stimuli. The inflammatory and other responses to lung and intestinal infections induce Fos protein expression in the key hindbrain signal-relay centres of the nucleus of the solitary tract and the lateral parabrachial nucleus, but the pattern of this activation can be distinguished according to the tissue localisation of the parasites. Nevertheless, both phases of anorexia are likely to be related to pathological events consequent upon the parasite burden, and it is possible that common or related signalling pathways will mediate these effects.

Feeding behaviour in the parasitised rat

Many studies of food intake in laboratory animals simply involve measuring daily food consumption by difference weighing (i.e. quantity of food at start minus quantity of food remaining after 24h). While this is clearly a very convenient experimental approach it is both rather crude and it involves disturbance of animals which itself may alter feeding behaviours. A more precise and less interventionist approach involves remote sensing to measure the quantity of food eaten as and when it is taken by using a feeding activity monitor. We have employed just such an approach to examine the precise patterns of feeding in rats during the course of infection with *N. brasiliensis.* The apparatus consisted of 14 modified rat boxes each housing a single animal with access to food gained through a series of holes cut in the floor of the box. Under these holes crushed pellet food was placed in a dish itself placed on a balance (Mettler PB302, $\pm$ 0.01g). Each of the 14 balances was connected by an individual RS232 lead to a computer containing the appropriate software (Feeding and Drinking Monitor v4.60, Columbus Instruments, Ohio, USA). The programme supported continuous monitoring of the quantity of food taken, the duration of feeding and the times of onset and termination of feeding bouts in 14 animals simultaneously (7 uninfected controls and 7 infected rats). Data were sampled and downloaded every 20 minutes over periods of between 11 and 20 days.

The results typically obtained from the continuous monitoring of food intake and feeding patterns are illustrated in Figure 2 which shows the feeding profile for an individual rat for 3 days prior to infection and for 8 days post-infection.

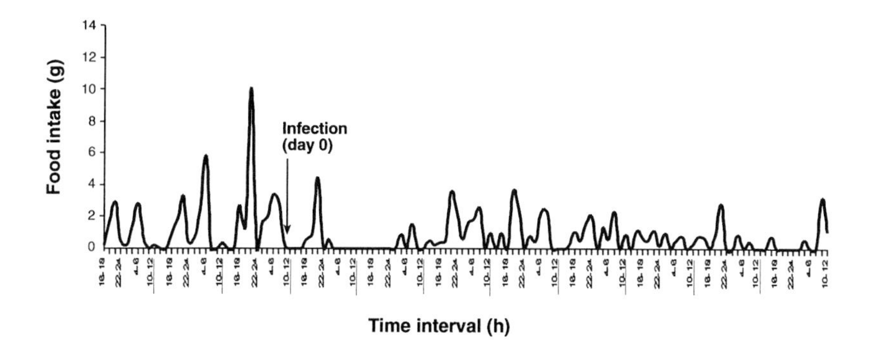

Figure 2. The feeding profile of a single rat infected with *Nippostrongylus brasiliensis* obtained from the continuous monitoring of food intake for 3 days prior to infection and for 8 days post-infection. The point of infection is indicated. Food intake was monitored every 20 minutes. Under a 12:12 light/dark cycle (light phase 08.00h to 20.00h) uninfected rats feed primarily during the dark phase with two discernible bouts of feeding activity. The normal feeding pattern was rapidly disrupted by infection and by 18h feeding had stopped completely.

Under standard laboratory conditions of 12:12 light/dark cycle (light phase 08.00h to 20.00h) uninfected rats fed primarily during the dark phase with two discernible bouts of feeding activity. Only minimal quantities of food were taken during the light phase usually in single feeding episodes. This pattern was maintained for less than 24h post-infection and by 18h feeding had either been severely reduced or stopped completely. In the example illustrated in Figure 2 the rat stopped feeding at 18h post-infection and resumed at 40h. In other animals in the same experiment some feeding occurred during this period but it was always minimal in terms of both quantity of food taken and the number and duration of feeding bouts. On resumption of feeding at around 40h post-infection the pre-infection biphasic pattern of nocturnal feeding was lost and animals fed sporadically both by day and by night. The first phase of anorexia therefore, commencing coincident with pulmonary invasion by parasite larvae, can be acute starting almost always 18h post-infection and lasting for 22 hours (Chappell, Geissler & Boyd, unpublished). It heralds a breakdown in the normal pattern of feeding that is sustained until around day 12.

The second phase of anorexia coincides with worm maturation in the small intestine and the onset of egg production; it is far less clearly defined than the lung phase commencing around day 6 post-infection and lasting until around day 9. During this period animals feed sporadically both by day and by night with repeated short feeding bouts. Following this phase there is a period of compensatory hyperphagia that varies in magnitude in individual animals. This hyperphagia is characterized by increases in intensity and duration of nocturnal feeding bouts.

Thus the application of a food intake monitor extends considerably the data obtained by measuring food intake on a once daily basis. It reveals the precise effects of *N. brasiliensis* infection on host feeding patterns indicating in particular the profound disruptive influence of the lung-phase anorexic episode as well as the exact time of its commencement. These data suggest that causal influences for this phase should be sought over a period between 16-18 hours post-infection. Furthermore, the two distinct phases of anorexia that occur in the infected rat are probably initiated and maintained by different pathophysiological events since their characteristics are quite different. These propositions now require further investigation.

Correlates of parasite induced anorexia

In mammals, the control of food intake and body weight involves the integration of numerous signals of neural, metabolic, endocrine and neuroendocrine origins (Schwartz *et al.*, 2000; Woods *et al.*, 1998). Despite the complexity of these integratory processes, the hierarchy and interactions of different components of the system have begun to emerge over the last decades. Thus signals that regulate food intake include both those that originate peripherally (i.e. outside the CNS), and those within CNS integratory centres. Peripheral signals relate to the consequences of meal processing, gastrointestinal activity and flux in energy stores, and have predominantly short duration effects. Drawing upon knowledge of peripheral feedback onto CNS integratory pathways we have examined relevant signalling systems at several different levels.

Blood leptin concentration

Leptin, the hormone product of the *obese* gene, is secreted by adipose tissue and circulates in the bloodstream at concentrations that are broadly proportional to the level of adiposity of the animal. Leptin enters the brain by crossing or bypassing the blood-brain barrier, and interacts with receptors on hypothalamic neurones with a number of different neurochemical phenotypes. Via this feedback loop, leptin provides information on peripheral energetic status to regulatory systems in the brain. There appear to be two distinct forms of feedback provided by the leptin signal. Profound diurnal changes in leptin gene expression and blood leptin concentration reflect the daily pattern of food intake; leptin rises 4-6 hours out of phase with food intake and in rodents peaks during the later part of the night. This pattern of leptin synthesis and secretion is sensitive to changes in feeding patterns, suggesting that it is food intake that controls its gene expression. Low leptin levels and the absence of this daily leptin surge are potent signals of starvation (Ahima *et al.*, 1996; Ahima & Flier 2000). When food supplies are not limiting, basal concentrations of leptin in the bloodstream during the daily (daytime) nadir encode for the levels of body fat storage. This latter role may be secondary in importance to the role of leptin as a food intake/starvation signal. In small mammals, the regulatory loops formed by leptin and its downstream signalling systems permit 'sensing' of both daily food intake and body fat stores by the hypothalamus, and thus expression of the behavioural and metabolic adjustments required to regulate energy balance.

With this background, parasite-induced anorexia might be expected to have a negative effect on leptin feedback in proportion to the severity of the anorexia at different stages of the infection. The availability of a sensitive assay for plasma or serum leptin facilitated measurement of this hormone in blood samples from infected rats at different stages of the infection and from animals pair-fed to the same level of food intake. The hormone was measured using a sensitive enzyme-linked immunosorbent assay (ELISA; Hardie *et al.*, 1996), employing an antiserum raised against recombinant mouse leptin, but which recognizes both rat and mouse proteins. The rat and mouse proteins have 96% homology at the amino acid level. The developed assay had a detection limit of approximately 100 pg leptin/ml.

Our first investigation of plasma leptin concentrations in rats infected with *N. brasiliensis* compared groups of rats at 1, 2, 4, 8 and 17 days post-infection with uninfected and pair-fed controls (Roberts *et al.*, 1999). Compared with uninfected controls, leptin levels were significantly reduced in both infected and pair-fed groups at days 1 and 8 post-infection. The same trend was observed on day 4 post-infection. There was no difference between infected and pair-fed groups at these time points. On day 17 post-infection, by which time normal levels of food intake had been restored, leptin levels were similar in all three groups. However, on day 2 of infection, co-incident with the first of the two anorexic phases, plasma leptin concentrations were elevated in the infected group relative to both uninfected controls and pair-fed controls. An intriguing aspect of the increased plasma leptin on day 2 of the infection is that the statistical difference between infected and uninfected rats was largely the result of very high hormone titres in individual animals, rather than more modest increases in concentration in the whole population (Roberts *et al.*, 1999). Thus the concentration of leptin in the plasma of infected rats generally reflected the prevailing state of negative energy balance, but was subject to marked perturbation in individuals where sporadic high titres were recorded. To investigate the possible involvement of leptin in the acute phase of infection, rats were infected at either dark onset or light onset, and experiments terminated 12 or 24 hours later. Significant changes in plasma leptin were only observed in the 12 hour dark onset experiment, where concentrations were again elevated in infected rats.

Interpretation of leptin concentrations during the anorexia induced by worm infection was complicated further by the results of subsequent studies. Whereas it initially appeared that leptin was sporadically elevated in the plasma of rats during the first 48 hours of *N. brasiliensis* infections i.e. the first phase of anorexia, a subsequent study extended this phenomenon to the second phase of anorexia. At 8 days post-infection, two of six infected rats had elevated plasma leptin, although there were no significant effects when group means were compared (Mercer *et al.*, 2000). Only rats with a patent continuing infection had elevated leptin. Prior to these findings it appeared that while acute increases in plasma leptin might be involved in the induction of anorexia, more chronic changes (i.e. depressed leptin concentration) were secondary to the anorexia and reflected the state of negative energy balance in parasitised and pair-fed rats. Although there are a number of arguments against the proposition that leptin directly induces and maintains anorexia during the infection (Roberts *et al.*, 1999), the discovery of high plasma leptin concentrations in at least some rats at 8 days post-infection provides some further support for this. Resolution of some of the questions that arise from these observations may require studies to determine whether plasma concentrations are chronically elevated in these individuals, or whether leptin secretion is pulsatile.

Blood insulin and corticosterone concentrations

Our earliest studies of the hormonal and neuroendocrine characteristics of *N. brasiliensis* infections of the laboratory rat revealed major differences in plasma insulin concentrations between study groups (Horbury *et al.*, 1995). On day 2 post-infection, insulin concentrations were lower in infected and pair-fed rats than in *ad libitum* fed uninfected controls. The same depression in plasma insulin was observed in infected and pair-fed rats on day 8 post-infection. These differences in insulin concentration were observed in animals with a cumulative anorexia (mean reduction in food intake) over 2 and 8 days of 72% and 37%, respectively. In a subsequent study when hormone levels were assayed at 2, 4, 8 and 17 days post-infection, trends towards lower insulin concentrations in infected and pair-fed rats were observed at 8 and 17 days, but not at 2 and 4 days post-infection (Roberts *et al.*, 1999). Indeed at 4 days post-infection, immediately prior to the initiation of the second anorexic episode, insulin concentrations in infected rats were elevated relative to pair-fed controls. This last observation was accompanied by perturbations to the adrenal steroid system; plasma corticosterone concentration was higher in pair-fed rats than in either uninfected *ad libitum* fed or infected groups. The continued presence of a patent infection appears to be a prerequisite for elevated plasma insulin (Mercer *et al.*, 2000).

Concentrations of the key metabolic hormones, insulin and corticosterone, are also perturbed in the blood of infected rats during the early stages of infection. Twelve-hour infections initiated at either dark onset or light onset were characterized by a substantial increase in plasma corticosterone relative to control and pair-fed rats (Roberts *et al.*, 1999). In 12 hour infections initiated at dark onset, where plasma leptin was elevated, there was also increased plasma insulin concentration in infected rats. Infections of 24 hours duration provided evidence of reduced concentrations of plasma insulin in pair-fed rats relative to infected and uninfected control groups. In these acute experiments, changes in plasma insulin levels were observed in the animal groups with the greatest proportional anorexia. Further investigation of the striking elevation of plasma insulin and corticosterone during the first 12 hours of infection may provide more insight into the induction of anorexia, but the more chronic changes appear to be secondary to the anorexia that is maintained until parasite expulsion. The effects on plasma hormones must be independent of the relative hypophagia that results from the parasitic infection, since these parameters are unaffected in pair-fed groups. In summary, perturbations to leptin, insulin and corticosterone signals early in infection may have a causative role in parasite-induced anorexia. All three hormones are capable of regulating hypothalamic neuropeptide gene expression. However, changes in these circulating endocrine factors in more chronic infections are likely to be secondary to negative energy balance.

Hypothalamic neuropeptide gene expression

Located at the base of the forebrain, the hypothalamus integrates an array of endocrine and neuroendocrine signals from various body organs and brain regions, and several hypothalamic areas are active in the maintenance of energy balance through regulation of food intake and energy expenditure. Hypothalamic neuropeptides concerned with energy homeostasis can be divided into two categories. The anabolic peptides, such as neuropeptide Y (NPY), increase food intake, reduce energy expenditure, and give rise to positive energy balance and weight gain, while catabolic peptides have the opposite effects on each side of the energy balance equation, leading to negative energy balance. For peptides such as NPY, the evidence for involvement in energy balance is compelling (Schwartz *et al.*, 2000; Woods *et al.*, 1998). Within the rat hypothalamus, NPY is synthesised mainly in the arcuate nucleus (ARC), from which axons project to the paraventricular nucleus (PVN). Negative energy balance, such as that induced by food deprivation or restriction, is accompanied by increases in NPY mRNA in the ARC, and NPY peptide in the ARC and PVN. Elevated activity of the NPYergic system reflects an increased drive to eat and leads to hyperphagia upon refeeding and restoration of normal body weight. Injected NPY stimulates food intake and reduces energy expenditure, even in satiated rats, and can induce obesity.

The technique of *in situ* hybridisation allows the expression of genes to be quantified within brain sections while retaining anatomical resolution, making this methodology particularly appropriate for studying the effect on neuronal signalling systems of manipulations that alter energy balance. We have investigated the involvement of NPY in the establishment and maintenance of anorexia in rats infected with *N. brasiliensis*, asking whether the gene expression of these neuropeptides is affected by the changes in food intake and body weight that accompany infection. On day 2 post-infection, NPY gene expression in the ARC was elevated in pair-fed rats compared to *ad libitum* fed uninfected controls, and there was a similar trend in rats infected with *N. brasiliensis*. By day 8 post-infection similar levels of elevated NPY gene expression were recorded in the ARC of infected and pair-fed rats with established anorexia or imposed hypophagia. Elevated expression of the gene for this anabolic neuropeptide thus reflected the state of negative energy balance (Horbury *et al.*, 1995). Both infected and pair-fed groups exhibited a correlation between degree of anorexia and induction of NPY mRNA. Similar analysis on day 16, during the compensatory hyperphagia that follows the expulsion of the parasite by the host immune response, indicated that NPY gene expression remained high during at least the early stages of weight rebound. Thus the anorexia induces an increase in NPYergic activity but the normal behavioural sequelae of this (i.e. increased food intake) are not observed. This suggests that animals detect an energy deficit during the early stages of the infection, yet do not feed, only becoming hyperphagic coincident with worm loss. The failure of anorexic parasitised animals to feed in response to activation of NPYergic pathways suggests that the NPY feeding stimulus may be blocked or overridden by other neural and/or endocrine signals activated by infection or possibly of parasite origin. Clearly, however, such a blockade of appetite drive could be a fundamental mechanism of relevance to both parasitism of humans and animals as well as eating disorders such as anorexia.

Cytokines

Infections with micro- or macro-parasites result in the production within the host of a battery of cytokines. These endogenously produced immunomodulatory molecules not only regulate immune responses to invading organisms but may also affect food intake, and thus may be causally linked to anorexia. Exogenously administered cytokines such as interleukin(IL)-1, IL-6, IL-8 and tumour necrosis factor (TNF) can lower food intake in laboratory rodents (Plata-Salaman, 1995). In the specific context of *N. brasiliensis* infections of the rat, IL-1, IL-6 and TNF are released during the lung phase of infection (around day 2 p.i.; Benbernou *et al.*, 1992; Fig. 3), and IL-6 is a known mediator of anorexia in acute phase protein responses (Moshyedi *et al.*, 1998).

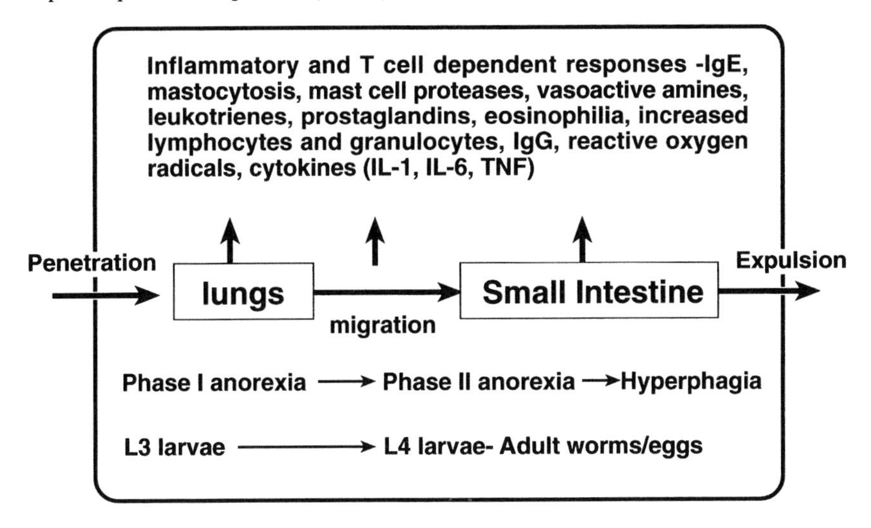

Figure 3. Events in the biology of *Nippostrongylus brasiliensis* associated with worm expulsion and changes in appetite. The complex specific and non-specific immunological mechanisms that lead to worm expulsion are not fully understood. Several of the resulting responses to infection could give rise to food intake changes.

It seems plausible therefore that parasitic infection may alter host appetite via cytokine mediation. In addition to the potential of peripheral cytokine synthesis to influence peripheral neural and endocrine signalling pathways, or to have a direct action on the brain, cytokine synthesis may also occur within the CNS itself. This is an area of active investigation in bacterial and microbial infections, but is as yet unexplored for helminth parasites. An additional intriguing possibility is that parasites themselves may produce cytokine analogues. For instance, *Trichuris trichiura* produces an interferon-γ-like molecule (Grencis & Entwistle, 1997) and all parasites excrete or secrete (E/S) products that interact with the host's immune system (see Cox, 2001). These E/S products may be capable of modulating the immune response and may, in addition, also mimic cytokine-like activities in other host compartments, such as appetite regulation.

Wasting disorders and the consequences of negative energy balance

The possibility that inflammation and cytokine signalling may be causally implicated in parasite-induced anorexia finds parallels in the study of human wasting disorders (Schwartz *et al.*, 1997). Sustained weight loss is a frequent symptom among patients with AIDS or cancer, raising questions similar to those posed earlier in this review; namely, why compensatory responses are unsuccessful in defending body weight under these circumstances. As outlined earlier, the hypothalamic compensatory systems rely on feedback provided by peripheral hormonal signals including leptin, insulin and glucocorticoids. Negative energy balance and weight loss is accompanied by reductions in plasma concentrations of insulin and leptin, and by increases in glucocorticoids. Reduced negative feedback then modulates the activity of anabolic and catabolic hypothalamic systems to co-ordinate changes in food intake and energy expenditure, thereby correcting the energy deficit. Arguments can be constructed to account for weight loss failing to produce appropriate behavioural and physiological changes. These centre on the disease process resulting in either inappropriate release of the signalling molecules in question, or the production of molecular mimics of these factors, thereby blocking the normal compensatory response. Cytokines may be produced in certain of the wasting illnesses and they could activate catabolic pathways within the brain. The possibility of induced cytokine release mimicking a high leptin signal can be rationalised from the assignment of the leptin receptor to the class-I cytokine receptor family (Tartaglia, 1997), and the similarities in signal transduction pathways between the receptors in this family. The implication of cytokines in the human wasting illnesses, reinforces the need to regard these molecules as possible mediators of parasite-induced anorexia.

The expression of substantial anorexia in parasitised animals, despite the provision of food in excess, appears likely to initiate the classical neuroendocrine responses to negative energy balance. This response conserves energy in order to optimise the use of limited reserves and prioritise towards vital functions, and includes suppression of the hypothalamo-pituitary-gonadal and -thyroid axes, and increased activity of the hypothalamo-pituitary-adrenal axis. The ability of leptin to blunt all these adaptations to starvation serves to emphasise the pivotal role of this molecule in signalling negative energy balance. If the responses outlined above can be interpreted as shutting down less essential systems, then the complex immune response system also appears to fall into this category. Malnutrition predisposes the individual to death or disease from infections, and starvation and low body weight suppresses immune function, with impaired cell-mediated immunity. Here again, leptin appears to form the link between nutritional status and cellular immune function. Leptin is able to reverse the immunosuppressive affect of acute starvation and its own synthesis is sensitive to (stimulated by) inflammatory cytokines. As discussed earlier, it is possible that a less severe but chronic negative energy balance may not be accompanied by the same physiological and endocrine responses. Indeed since nutrient overloads may have immunotoxic effects, it is not inconceivable that a mild anorexia could have a positive impact on the immune response.

The parasite is required for parasite-induced anorexia

We have investigated those biological events within the host-parasite relationship which are critical to the development of each of the two distinct anorexic phases that characterize *N. brasiliensis* infections in the laboratory rat. Using the anthelminthic, mebendazole, *N. brasiliensis* infections were eliminated between the first and second anorexic episodes (Mercer *et al.*, 2000). This intervention prevented the expression of the second phase of anorexia (Fig. 4).

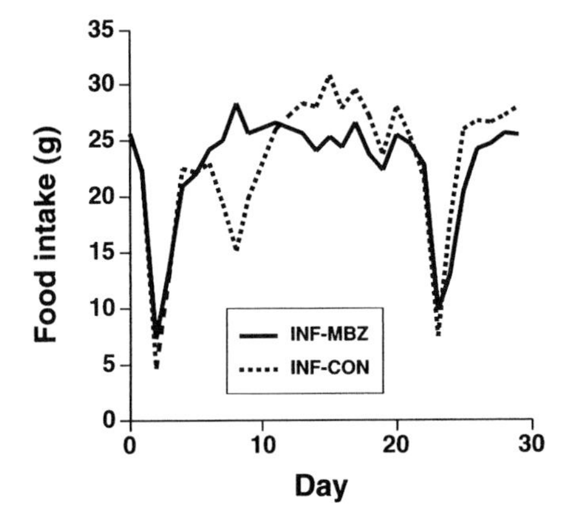

Figure 4. Daily food intake (g per day) of rats infected with *Nippostrongylus brasiliensis* on day 0 and given a secondary challenge infection on day 21. One of the groups (INF+MBZ) was treated with the anthelminthic, mebendazole, on days 2 and 4, thereby eliminating the second anorexic episode. Only a single anorectic episode was observed in secondary infections.

Rats exposed to a second infection with *N. brasiliensis,* three weeks after the primary infection, exhibited only a first phase anorexic response, the magnitude of which was similar to that observed in the primary infection and which was not influenced by mebendazole termination of the primary infection. Inoculation of rats with heat-killed *N. brasiliensis* larvae failed to induce anorexia and did not alter the severity of biphasic anorexia on subsequent injection of viable larvae. The continuing presence of live parasites in the appropriate body location, i. e. lungs or small intestines, therefore appears to be a prerequisite for ongoing biphasic anorexia Thus the first anorexic episode is dependent upon viable migrating larvae, while the second clearly requires the continuing presence of a patent parasite infection beyond the lung phase. The importance of a continuing infection in the maintenance of anorexia finds parallels in studies of sheep with long-term trickle infections of *Trichostrongylus colubriformis*. Ivermectin treatment of sheep with chronic monophasic parasite-induced anorexia rapidly restored food intake to levels of uninfected control animals (Kyriazakis *et al.*, 1996).

Concluding remarks

These parasite studies are in the early stages and much remains to be unravelled concerning the mechanisms underlying parasite-induced anorexia. Many of the modulatory molecules that are very recent arrivals on the appetite/energy balance scene have still to be examined in this context. However, based on our studies of NPY, it might be anticipated that any modulation in the activity of these other hypothalamic neuropeptide systems could also be secondary to and thus simply reflect the state of negative energy balance. The cytokines, including leptin, remain as candidate mediators of the onset and maintenance of anorexia and further studies of these systems are merited. Another intriguing possibility is that secreted products, originating in the parasite itself, could affect the food intake of the host. It has been suggested, for example, that parasites synthesise and secrete cytokine analogues, and at least one worm parasite produces an analogue of mammalian growth hormone. With the evolution of molecular approaches to these problems we are now in a position to tackle such issues. Parasites affect many millions of people and animals worldwide, thereby causing both considerable suffering and significant economic losses. The more we discover about the basic biology of parasites and the ways in which they manipulate their hosts, directly or indirectly, the closer we may come to resolving some of these problems. The occurrence of similar endocrine and neuroendocrine 'symptoms' during the two phases of anorexia in the rat suggests that this laboratory model should provide valuable insight into parasite-induced anorexia in commercially-important species.

Acknowledgements

This work was funded by grants from The Leverhulme Trust (F/152/N), BBSRC, Wellcome Trust, and Rank Prize Funds and with support from the Scottish Executive Rural Affairs Department.

REFERENCES

Ahima RS, Prabakaran D, Mantzoros C, Qu D, Lowell B, Maratos-Flier E, Flier JS. Role of leptin in the neuroendocrine response to fasting. *Nature* 1996; 382:250-52

Ahima RS, Flier JS. Leptin. *Ann. Rev. Physiol.* 2000; 25:413-27

Benbernou N, Matsiota BP, Jolivet C, Ougen P, Guenounou M. Tumour necrosis factor, interleukin-1 and interleukin-6 in bronchoalveolar washings and their *in vitro* production during *Nippostrongylus brasiliensis* infection. *Clin Exp Immunol* 1992; 88:264-68

Castex N, Fioramonti J, Ducos De Lahitte J, Luffau G, More J, Bueno L. Brain Fos expression and intestinal motor alterations during nematode-induced inflammation in the rat. *Am J Physiol* 1998; 274:G210-16

Coop RL, Holmes PH. Nutrition and parasite interaction. *Int J Parasitol* 1996; 26:951-62

Cox FEG. Concomitant infections. Parasites and immune responses. *Parasitology* 2001; 122 (in press)

Exton MS. Infection-induced anorexia: active host defence strategy. *Appetite* 1997; 29:369-83

Fox MT. Pathophysiology of infection with gastrointestinal nematodes in domestic ruminants: recent developments. *Vet Parasitol* 1997; 72:285-97

Grencis RK, Entwistle GM. Production of an interferon-gamma homologue by an intestinal nematode: functionally significant or interesting artefact? *Parasitology* 1997; 115:S101-S105

Hardie LJ, Rayner DV, Holmes S, Trayhurn P. Circulating leptin levels are modulated by fasting, cold exposure and insulin administration in lean but not Zucker (*fa/fa*) rats as measured by ELISA. *Biochem Biophys Res Commun* 1996; 223:660-65

Horbury SR, Mercer JG, Chappell LH. Anorexia induced by the parasitic nematode, *Nippostrongylus brasiliensis*: effects on NPY and CRF gene expression in the rat hypothalamus. *J Neuroendocrinol* 1995; 7:867-73

Kyriazakis I, Tolkamp BJ, Hutchings MR. Towards a functional explanation for the occurrence of anorexia during parasitic infections. *Anim Behav* 1998; 56:265-74

Kyriazakis I, Anderson DH, Oldham JD, Coop RL, Jackson F. Long-term subclinical infection with Trichostrongylus colubriformis: effects on food intake, diet selection and performance of growing lambs. *Vet Parasitol* 1996; 61:297-313

Langhans W. Anorexia of infection: current prospects. *Nutrition* 2000; 16:996-1005

Mercer JG, Mitchell PI, Moar KM, Bissett A, Geissler S, Bruce K, Chapell LH. Anorexia in rats infected with the nematode, *Nippostrongylus brasiliensis*: experimental manipulations. *Parasitology* 2000; 120:641-47

Moshyedi AK, Josephs MD, Abdalla EK, Mackay SLD, Edwards CK, Copeland EM, Moldawer LL. Increased leptin expression in mice with bacterial peritonitis is partially regulated by tumor necrosis factor alpha. *Infect Immun* 1998; 66:1800-02

Ovington KS. Dose-dependent relationships between *Nippostrongylus brasiliensis* populations and rat food intake. *Parasitology* 1985; 91:157-67

Plata-Salaman CR. Cytokines and feeding suppression: an integrative view from neurologic to molecular level. *Nutrition* 1995; 11:674-677

Roberts HC, Hardie LJ, Chappell LH, Mercer JG. Parasite-induced anorexia: leptin, insulin and corticosterone responses to infection with the nematode, *Nippostrongylus brasiliensis*. *Parasitology* 1999; 118:117-23

Schwartz MW, Seeley RJ, Woods SC. Wasting illness as a disorder of body weight regulation. *Proc Nutr Soc* 1997; 56:785-91

Schwartz MW, Woods SC, Porte D Jr, Seeley RJ, Baskin DG. Central nervous system control of food intake. *Nature* 2000; 404:661-671

Tartaglia LA. The leptin receptor. *J Biol Chem* 1997; 272:6093-96

Woods SC, Seeley RJ, Porte D, Schwartz MW. Signals that regulate food intake and energy homeostasis. *Science* 1998; 280:1378-83

CONCLUSION

Implications for understanding and treating human eating disorders

Conclusion

Human eating disorders may best be regarded as forming a spectrum from the low food intake and emaciation of classical restricting anorexia nervosa through bulimia and bingeing to hyper-obesity (Chapter 2). The border between these points on the spectrum may often be blurred as may their distinction from 'normal' leanness and 'normal' heavy build expected in 'normal' biological variation.

Animals in the wild appear to be relatively free from eating disorders although they are subject to fluctuations in body composition according to seasonal and reproductive cycle factors (Chapter 4). However domesticated animals, including farm, laboratory and pet animals, are prone to disorders similar to some of those on the human spectrum.

The book considers the two main aspects of input/output involved in homeostasis of body composition and of disorders in its operation:

1) Appetite - including voluntary choice of both diet composition and level of consumption

2) Metabolic output - including voluntary activity (exercise)

Both of these are a composite of traits that are influenced by genetic and non-genetic factors. In addition correlation between the component traits can be particularly important in any attempt at preventing and/or treating disorders.

Several relevant points are covered in this book:

- The validity and usefulness of considering animal models in relation to the spectrum of human eating disorders. Chapter 3 considers animal welfare objections and indicates facets of human disorders that can invalidate the use of animal models. In particular economic considerations - in the broad sense of decision-making in eating and lifestyle - are largely distinctive human preoccupations.

- Much of the book focuses on the physiological and underlying genetic basis of appetite and metabolism that appear to be common to the higher animals. Cross-species study of these aspects seems to be a fruitful joint enterprise even if the studies are non-interventionist observations of wild and domesticated species. Animal models are shown to have contributed significantly to the fund of knowledge underlying our understanding of eating behaviour and body composition homeostasis. On the genetic side the flow of information is both ways. The human genome project is of great value in filling gaps in the genome maps of the other species. Animal models for their part have been useful in suggesting possible candidate genes underlying specific disorders e.g the *ob* gene and its product leptin. They are also making contributions to knowledge of gene effects by means of 'knockout gene' experimentation. Although such procedures are difficult to execute and to interpret, they are promising additions to the tools relevant to eating disorders, provided animal welfare is properly taken into account.

- The animal models have also been valuable in addressing the vexed question of the correlation of eating/body composition disorders and the two major environmental influences inviolved - diet choice and voluntary activity (Chapters 4,5,6). In terms of treatment and prevention the genetic predilection of subjects genetically susceptible to obesity or anorexia to choose diets and activity patterns inimical to their welfare makes things difficult allround.

- Animal models confirm that both genetic and environmental factors are involved in eating and body composition disorders in a complex way. It suggests that human sufferers can exert less will-directed action to help their cause. May not a more realistic appreciation of this not only help the patient in terms of a clearer conscience and personal dignity but engender a more realistictic view of prevention and treatment on behalf of the medical adviser?

Subject Index